9 E.T.H.E.

# 9
# E.T.H.E.R.
# R.E. ENGINEERING

By

African Creation Energy

June 26, 2012

www.AfricanCreationEnergy.com

# R.E. ENGINEERING

*All Rights Reserved*

Copyright © 2012 by African Creation Energy,
Written and Illustrated by African Creation Energy,
www.AfricanCreationEnergy.com
No part of this book may be reproduced or transmitted in any form or by any means, electronic or mechanical, including photocopying, recording, or by any information storage and retrieval system without permission in writing from the author.

**ISBN 978-1-105-82436-4**

Printed in the United States of America

# 9 E.T.H.E.R.

*"For, behold, the day cometh, that shall* **Burn as an Oven**...*and the day that cometh shall* **Burn them up**...*but unto you that fear my name shall the* **ETHER** (SUN *of Righteousness*) *arise with* **Healing in his Wings**..."
~ ***Bible Old Testament, Book of Malachi 4:1-2***

*"John answered saying, 'I baptize you with* **Water**; *but someone more* **Powerful** *than I is coming...and he shall Baptize you with* **ETHER** *(the Holy* SPIRIT *and* FIRE*)"*
~ ***Bible New Testament, Gospel of Luke 3:16***

*"Allah is* **ETHER** *(the* LIGHT *of the* HEAVENS *and the Earth...*LIGHT *upon* LIGHT*)! Allah doth guide whom He will to His* **Light**, *Allah doth set forth Parables for men, and Allah does* **know all things**."
~***Quran, Surah An-Nur 24:35***

# R.E. ENGINEERING

### AFRICAN CREATION ENERGY ALERT:

In Physics, "**Coherent Radiation**" is defined as the Orderly emission of Energy from a source. Radiation is considered Coherent when it does not diverge and the frequency phase difference is constant. Coherent Radiation enables the transmission of Frequencies with low levels of loss. The symbol above is a modification of the warning symbol for Coherent Radiation. Just like Coherent Radiation permits Frequencies to be transmitted with very little loss; in order for Information to be comprehended, it must be transmitted Coherently or Orderly. This book deals with the concept of Ether or Æther, which is an Ancient concept whose information has become somewhat lost in transmission from the past to the present, but has practical implications and applications in modern Theology, Theory, and Technology. Thus, this book is designed to coherently transmit quintessential information about the many applications of Ether or Æther. Therefore, African Creation Energy's modified version of the "Coherent Radiation" warning symbol appears here as a sign to signify the Coherent, Harmonious, and Orderly "Etheric" frequencies that are emitted within these pages and absorbed by the reader if the Etheric Information within is tested, experienced, and applied. **Proceeded if you Will.**

African Creation Energy

Creative Solution-Based Technical Consulting

# 9 E.T.H.E.R.

## Table of Contents

| Section | Page |
|---|---|
| 1.0. ÆTHER RE-INTRODUCTION | 6 |
| 2.0. ANTIMATTER | 33 |
| 3.0. ELECTRICITY | 38 |
| 4.0. THERMODYNAMICS | 41 |
| 5.0. HYDRODYNAMICS | 50 |
| 6.0. ELECTROMAGNETIC RADIATION | 59 |
| 7.0. RESONANT ENERGY | 78 |
| 8.0. ETHER EXPERIMENTS TO EXPERIENCE EVIDENCE... | 81 |
|     Calculate the Antimatter emitted from an African Yam | 81 |
|     Create a Staff of PTAH to convert DC to AC | 82 |
|     Create a Solar Barque of RE (Pop-Pop Boat) | 83 |
|     Create a Hot-Air Balloon | 84 |
|     Create a Shu Ankh (Aeolia Pile) | 85 |
|     Separate Water into Hydrogen and Oxygen via Electrolysis | 86 |
|     Create "Naphtha" Biodiesel Fuel | 87 |
|     Create a Black Light Candle | 88 |
|     Ankh Wireless Energy Transfer Resonant Transformer | 89 |
| 9.0. AFRICAN CREATION ENERGY ETHER | 90 |
| REferences and REsources | 99 |

# R.E. ENGINEERING

## 1.0. ÆTHER RE-INTRODUCTION

Since time immemorial, Humanity has acknowledged the existence of a fundamental and **Quintessential** element and energy that permeates the totality of Nature. While the names, labels, stories, details, and explanations regarding this quintessential substance has changed overtime, the concept has been present and prevalent in a variety of different cultural expressions, philosophies, theologies, ideologies, religions, theories, and sciences throughout history. In the English language, this Quintessential substance is called **Ether**, **Aether**, or **Æther** which comes from the Greek word "aithēr" meaning "*pure air*" or "*clear sky*". The word Ether is also related to the Greek word "**Aithio**" meaning "***to incinerate and burn***" which is found in the word **Aithiopian** or **Ethiopian** meaning "***Ether People***" which is a name the Greeks once used for all Africans. In Ancient Alchemy, Ether was conceptualized as the **Fifth element** after the four elements of **Earth, Air, Fire, and Water**. The word "Quintessential", which means "the essential essence of something", actually breaks down etymologically as "**Quint-**" meaning "**Fifth**" and "**-essential**" meaning "Element" rendering the Alchemical term "**Fifth Element**" which signifies the fundamental importance attributed to Ether. In Greek mythology, Aether was personified as a **Primordial Elemental** god representing "**Light**" and the **Upper Air** and **Upper Atmosphere** above the terrestrial sphere where the gods **lived** and **breathed**. The personification of Aether in Greek mythology was actually predated by Ancient African concepts, cosmologies, and cosmogonies which personified the concepts related to Ether through various anthropomorphic deities. In Ancient Egypt, the **Primordial** gods called the **8 Ogdoad** or

## 9 E.T.H.E.R.

**Khemenu** represented the concepts of Nothingness, Chaos, Void, Darkness, Infinity, Eternity, Hidden, and Invisible qualities in Nature. In Egyptian mythology, the **Ogdoad** or **Khemenu** gave birth to the **9 Ennead** or **Sedjet** which included deities that personified concepts related to Ether including **Atum** or **Re (Energy or Light)** and **Nut (Sky)** as well as deities which personified the other four Alchemical elements including **Shu (Air)**, **Tefnut (Water)**, **Geb (Earth)**, and **Sutekh (Fire)**.

Just as the Ancient Egyptians personified the various Energies, Elements, and concepts related to Ether, we find similar examples of this practice over time throughout Africa. Amongst the **Yoruba** people in present day **Nigeria**, the word **"ASHE"** is used to refer to the **Ethereal Power** and **Energy** which **causes all events and happenings**. The various forms of "Ashe" or "Ether" are personified in the pantheon of **Orisha** deities. The word "Orisha", breaks down into **"Ori-"** meaning "Head, Ruler, or Master" and **"-isha"** short for **"Ashe"** meaning **"Energy** or **Ether"**, and thus the word "Orisha" literally means **"Rulers or Masters of Ether"**. Some of the names for the Yoruba Orisha who are the personifications of various forms and concepts related to Ether are: **Olorun**, who is the **Ruler of the Heavens**, divine creator, and **source of all Energy/Ether**; **Oya Iansan** is the Orisha of the **winds**, **tornados**, and **hurricanes** who is the **"mother of nine children"** and wife to the Orisha of **Lightning** named **Shango**. To the **Fulani** people of West Africa, **Nun-fari** is the deity considered the Ancestor of all **Blacksmiths** who mastered **Ether** in the form of three **"spiritual fires"**, namely: **Wood Fire**, **Earth Fire**, and **Heavenly Fire**. The deities of the **Akan** people in **Ghana** are another example of the personification of the concepts related to Ether in Africa. Some of the names for the personifications of Ether in the Akan culture include **Nyame**, the **Sky deity**, whose **right eye** is said to be the **Sun** and whose **left eye** is said to be the **moon**. Variations of the Akan aspect of Ether called **Nyame** is also acknowledge by the Mende people in West Africa who call it **Nyama** and the Dogon people of West Africa who call it **Amma**.

In the Akan culture, Nyame's is said to be the father of **Anansi the Spider** who spins the **Web of Energy/Ether** that exists throughout all of existence. The Akan people of West Africa also acknowledge what they call the **Abosam**, which are various aspects of Ether/Energy that can be contained within assorted persons, places, or things.

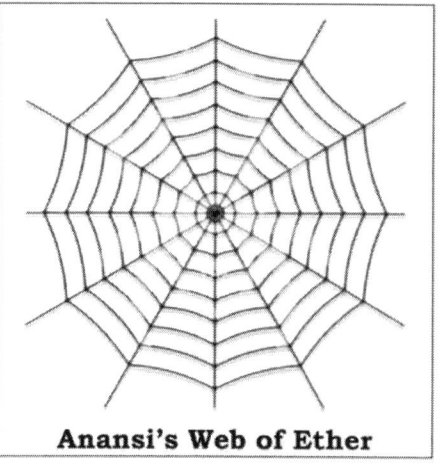

**Anansi's Web of Ether**

There are many other examples of Ether, and the deification of the concepts related to Ether, that can be found around the world. The Alchemical systems of the **Chinese**, **Japanese**, and **Tibetan** cultures all acknowledge ether. In **Hinduism**, Ether or Aether was called **AKASHA** and referred to "**Space**", the **cosmos**, or the "**Sky**". In Ancient **Mesopotamian** culture, concepts related to Ether were deified by the gods **Anshar**, god of the **sky**, **Anu**, god of the **Heavens and Constellations**, and **Enlil**, **god of breath and wind**. If charted cyclically, the various personifications or transformations of Ether/Energy from one form or personality to another is comparable to what modern Science now calls **Thermodynamics**. The thermodynamic processes of Ether were charted in the Ancient science of Alchemy in cultures around the world. In Africa in the Congo, Ether Energy is called "**Dikenga**" and the Thermodynamic or cyclic change in the various aspects of Dikenga is depicted in the Congo cosmogram called the "Yowa".

## 9 E.T.H.E.R.

In modern science, the acceptance of the concepts related to Ether has undergone a metaphorical thermodynamic change over time. Since modern Scientific Theory grew out of the Ancient Theologies, initially many of the early modern Scientist accepted and acknowledged the existence of Ether. **Plato, Aristotle**, and **Isaac Newton** all acknowledged and discussed Ether or Aether. Aristotle suggested that the **"highest heaven"**, called **Empyrean** (meaning **"in the fire"**), was filled with the element of Aether.

Early mathematicians equated each fundamental Alchemical element with a different geometric shape such that the Cube represented Earth, the Icosahedron represented Water, the Octahedron represented Air, the Tetrahedron represented Fire, and the Dodecahedron represented Ether. To most early modern scientist, Ether was acknowledged as a medium, field, or substance which filled space and matter and was necessary for the transmission and propagation of Light. This "light-bearing" ether was called **Luminiferous Aether**.

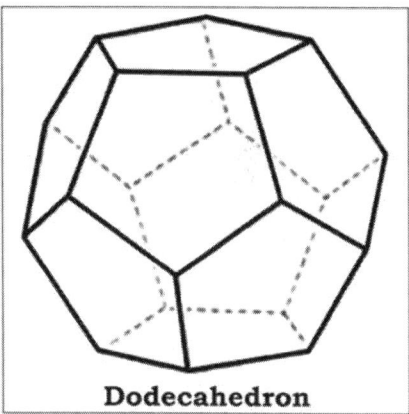
**Dodecahedron**

In the 1800s, various scientists tried to prove the existence of Ether experimentally, and were unsuccessful. By the early 1900s, Albert Einstein's theories of Relativity convinced most scientists at the time that Ether was completely imaginary and did not exist. However, the death of the early scientific concepts related to ether gave rise and rebirth to new modern scientific concepts which echo the concepts related to Ether of the past. Albert Einstein's theories which suggested that Space was some type of "medium" actually introduced what is called **Relativistic Ether** and the "**Einstein Æther Theory**". Einstein's Theory, along with more recent modern experiments, show that space is not an empty vacuum filled with "nothingness", but rather space is filled with a transparent substance with a structure comparable to solids, liquids, and gases. In addition to the scientific evidence that supports the existence of Relativistic Ether permeating space, other modern scientific concepts that are reminiscent of Ancient Ether concepts are **String Theory**, the **Higgs Field**, **Dark Energy**, and the **Grand Unified Theory of Everything**. While all of these advanced modern Scientific Theories include the idea of an all prevailing substance or material in the universe similar to the ancient and classical concept of Aether, empirical evidence of Ether can be obtained by simply observing evidence in Nature and comparing the observations to the qualities, characteristics, and definitions attributed to Ether from antiquity. From the Ancient and Classical definitions of Ether, there are basically **9** attributes that are consistently associated with Ether.

## 9 Attributes of Æther from Ancient Science are:

1) Æther permeates all of Nature, the Universe, and Existence

2) Æther is Primordial, Essential, and Fundamental

3) Æther exists in outer Space, the Sky, and Heaven

4) Æther exists within the Upper Air and Atmosphere

5) Æther is related to Breath and Breathing

6) Æther is needed for Life

7) Æther can Incinerate and Burn

8) Æther is related to Light

9) Æther is associated with Energy

## 9 E.T.H.E.R.

A comparative analysis of these 9 attributes of Æther from Ancient Theology and Ancient Science to basic modern scientific concepts and modern Religious concepts reveals striking similarities which will be expounded on below using a **Question-and-Answer** format. The Question-and-Answer format is being utilized to make these comparisons between Ancient Æther concepts to modern Æther concepts because the Question-and-Answer format is similar to the **"Call and Response"** format prevalent throughout African culture, and the Question-and-Answer format is also similar to the **"Transmitter and Receiver"** paradigm found throughout communication electronics which have both proven to be effective information conveyance modalities.

**1) Question: What are some examples of modern scientific concepts which are similar to the Ancient concept of Æther permeating all of Nature, the Universe, and Existence?**

**Answer:** In modern Physics, the most common form of matter in the Universe is **Plasma**. Plasma is a state of matter which results from the **Ionization** of **Gas**. Ionization occurs when the number of **Electrons** in an atom does not equal the number of **Protons** in an atom. Contrary to popular belief, outer Space is not a completely empty void vacuum, but rather filled with **Plasma**, **Gas**, **Electromagnetic Radiation**, and **Neutrinos**. The Electrons that Ionize gas and form Plasma are Fundamental Particles in Nature and are also used to form **Matter**. The Electromagnetic Radiation that permeates outer space is a fundamental force in Nature and motivates the movement of matter. Considering that Plasma, Electrons, Neutrinos, and Electromagnetic Radiation permeate all of Nature and the Universe, then these concepts can be considered analogous to **Æther**.

www.AfricanCreationEnergy.com

# R.E. ENGINEERING

**2) Question: What are some examples of modern scientific concepts which are similar to the Ancient concept of Æther being Primordial, Essential, and Fundamental?**

**Answer:** The Essential and Fundamental particles in Nature known at this time in modern science are **Leptons** (including **Electrons, Muons, Tauons,** and **Neutrinos**) and **Quarks**. The Primordial, Essential, and Fundamental Forces in Nature known at this time in modern science are the **Strong Force**, the **Weak Force, Electromagnetic Radiation**, and **Gravity**. In modern science the **Anti-particles** of the fundamental particles which combine to form **Antimatter** could be considered "Primordial". Thus, considering that Leptons (including Electrons, Muons, Tauons, and Neutrinos), Quarks, Antiparticles, Antimatter, the Strong Force, the Weak Force, Electromagnetic Radiation, and Gravity are all Primordial, Essential, and Fundamental concepts in modern science, then these concepts can also be considered analogous to **Æther**.

**3) Question: What are some examples of modern scientific concepts which are similar to the Ancient concept of Æther existing in outer Space, the Sky, and Heaven?**

**Answer:** As we previously stated, outer Space is not a completely void vacuum, but rather filled with Plasma, Gas, Electromagnetic Radiation, and Neutrinos. What makes outer Space a vacuum is that the pressure in outer space is much less than the atmospheric pressure on Earth. However, the Plasma, Gas, and Electromagnetic Radiation from outer Space also exist within the Heavens, Sky, and atmosphere on Earth. The plasma and gas that fills outer Space can combine to form **Stars** and **Solar Systems**. The **Nebular Hypothesis** in modern science states that the Solar System started as a **Molecular Cloud** of Interstellar **Gas** filled with particles of **all the existing gases and elements in Nature**. When these Molecular Clouds

## 9 E.T.H.E.R.

of Gas collapsed, they formed **Nebulas** and **Proto-planetary discs** which eventually became Stars, Planets, and Solar Systems. The **Orion Nebula** is a well known example of a Nebula that is **the combination of all existing gases in Nature** and is considered a **Star Nursery**. Thus, considering that Plasma, Gas, Electromagnetic Radiation, and Neutrinos fill outer Space, the Sky, and Heaven, then these modern scientific concepts can be considered analogous to the Ancient concept of **Æther**.

**4) Question: What are some examples of modern scientific concepts which are similar to the Ancient concept of Æther existing within the Upper Air and Atmosphere?**

**Answer:** As we previously mentioned, the Plasma, Gas, and Electromagnetic

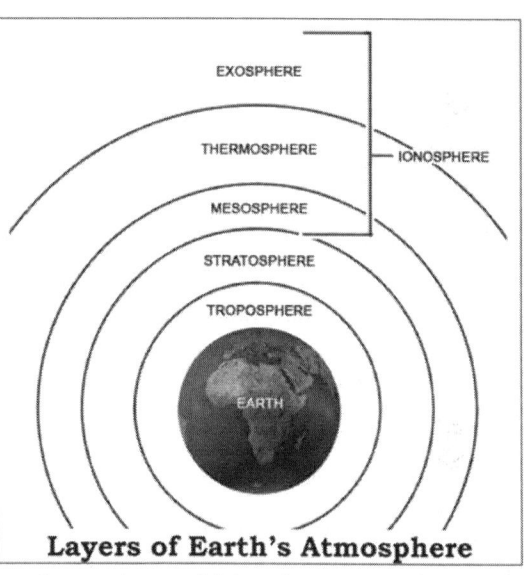

**Layers of Earth's Atmosphere**

Radiation from outer Space also exists within the atmosphere of Earth. In particular, in modern science the **Upper Atmosphere** of Earth is called the **Ionosphere** which consists of **Mesosphere**, **Thermosphere**, and **Exosphere**. Within the Thermosphere of the Ionosphere of the Upper Atmosphere of Earth, **Ultra-Violet Electromagnetic Radiation** causes **Ionization** and **Aurora light** displays in the sky. The atmosphere of Earth is composed of **Oxygen, Nitrogen, Carbon Dioxide, Hydrogen**, and the **Noble Gases** of **Argon, Neon, Helium, Krypton**, and **Xenon**. Thus, the ionized gas in the Ionosphere and Thermosphere along with

## R.E. ENGINEERING

Electromagnetic Radiation and the elemental components of the atmosphere can all be considered empirical evidence of the Ancient concept of **Æther**.

| Composition of Air on Planet Earth |||| 
|---|---|---|---|
| Gas | Ratio compared to Dry Air (%) || Chemical Symbol |
|  | By volume | By weight |  |
| Oxygen | 20.95 | 23.2 | O2 |
| Nitrogen | 78.09 | 75.47 | N2 |
| Carbon Dioxide | 0.03 | 0.046 | CO2 |
| Hydrogen | 0.00005 | ~0 | H2 |
| Argon * | 0.933 | 1.28 | Ar |
| Neon * | 0.0018 | 0.0012 | Ne |
| Helium * | 0.0005 | 0.00007 | He |
| Krypton * | 0.0001 | 0.0003 | Kr |
| Xenon * | $9 \times 10^{-6}$ | 0.00004 | Xe |
| * denotes a Noble Gas ||||

In the three main modern monotheistic religions, namely Judaism, Christianity, and Islam, the Upper Air or Upper Atmosphere is synonymous with "Heaven". Thus, the **Etheric Realm** or "**Astral Plane**" of the Upper Atmosphere is analogous to the concepts of the "**Kingdom of Heaven**", Paradise, and "**Jannah**" found in the religions of Judaism, Christianity, and Islam. In many Ancient cultures, the various deities that were used to personify the concepts of Ether were often associated with **birds** because birds were able to fly into the upper atmosphere or Ether. Birds are able to travel to altitudes higher in the Ethers than any other mammals because of bird's ability to extract large amounts of **oxygen** from the air. One of the most famous Ancient deities related to Ether, and represented by a bird, was the sun god **RE** or **HERU** who was depicted as a **Hawk** or **Falcon**. Hawks have the ability to fly 150 meters into the upper atmosphere. However, the bird that can fly the farthest into the Ethers or upper atmosphere is the **Bar Headed Goose** who can fly to heights of 8,850 meters.

**HERU**
aka HU-RE, representing the Sun Ether.
Ur "Fire" and Heru or "Hur" a "Fire bird" Ether Deity

Another important bird deity from Ancient culture is the **Phoenix** also known as the **Bennu** bird in Egypt. The Bennu bird was considered the **soul** of the sun god **RE** and the **Light** at the apex of the Pyramid. The name "Bennu" means **"to shine"**. The Bennu bird was equivalent to the Phoenix bird that would destroy itself in **flames** every 1000 years by **burning** itself and its nest to ashes only to rise from the ashes **reborn** and **resurrected**. The ability to travel into the "Ethers" or upper atmosphere is why **Angels** in religious traditions are often depicted with wings. Other creatures with the ability to travel into the Ethers are **bumblebees** which can fly at heights of 9,000 meters, and **butterflies** which can fly at altitudes of 6,000 meters.

# R.E. ENGINEERING

**5) Question: How is Æther related to Breath and Breathing?**

**Answer:** It has been established that Æther exists within the Atmosphere of the Air we breathe, thus Æther also enters our bodies through breathing. The breathing or **Respiration** process within our body breaks down the air we breathe to deliver **Energy** to our body and **brain**. There is also a Religious relationship between **Ether** and **Breath** and **Spirit** and **Soul**. The etymological and original meaning of the word **"Spirit"** is **"to breathe or blow"**. In the three main monotheistic religions (Judaism, Christianity, Islam), the Hebrew word translated as **"Spirit"** in the Old Testament of the Bible is **"Ruwach"** ( ) Strong's H7307) which means **"breath, wind**, or **air"**. The Hebrew word **"Ruwach"** ( ) is also related to the Arabic word **"Ruh"** which means **"spirit"** and is used as an attribute of the **Angel Gabriel** in Islam as **"al-Ruh al-Quds"** which means "the **Holy Spirit"** in Arabic. In the New Testament of the Bible, the word translated as **"Spirit"** is **Pneuma** (Strong's G4151) meaning **"breath"** in Greek. The Hebrew word translated as **"Soul"** in the Old Testament of the Bible is **"Nephesh"** ( ) Strong's H5315) coming from the root word **"Naphash"** ( ) Strong's H5314) meaning **"to take a breath"**. Since we comprehend that Æther is related to Gas, Air, and Breathing, and all of the original meanings of the words for Spirit and Soul meant breathing or air, then we must also conclude that Æther is related to Spirit and Soul.

Æther is also related to the concepts of Spirit and Soul found other religious traditions besides Judaism, Christianity, and Islam. The Hebrew word **"Nephesh"** meaning **"to take a breath"** is also related to the Ancient **Akkadian** word **"Nappahu"** meaning **"blowpipe"** and the Ancient Egyptian word

## 9 E.T.H.E.R.

"**Nefu**" meaning "**wind**, **air**, and **breath**". The Ancient Egyptians divided the Ether that entered into one's body using various aspects which together made the human soul including the **Akh** (meaning "**to shine or radiate light**" – the **Energetic Ether** within the body), the **Ba** (mentality or personality), **Sekhem** (**vital power, spark of life**), and the **Ka** (**breath of Life**, vital essence). The **Ka** from Ancient Egypt is also similar to the concept of **Ki**, **Qi**, or **Chi** from Chinese philosophy meaning **gas**, **breath of life**, **consciousness**, **vital energy**, and **life force**. In Chinese philosophy, it is said that the Ether Ki can be channeled through the palms for healing in a process called **Reiki**. The concept of **Ki** in China is also similar to the concept of **Prana** meaning "**vital breath of life**" in Hindu philosophy and all relate to Air, breath, spirit, soul, and Æther.

### 6) Question: How is Æther is needed for Life?

**Answer:** It has been established that Æther is in the air we breathe, and it is a well known fact that air is needed to sustain life, thus Æther is needed for Life. The Æther energy we get from the air we breathe enables our minds to be conscious, and as we have previously shown, Æther, air and breath are all related to the Religious concepts of Spirit and Soul which religious theology teaches is needed for Life. When asked to substantiate the existence of Spirit and Soul, overwhelmingly, believers in these concepts ultimate equate Spirit and Soul to Energy. Thus, the study of Energy is the only substantial and verifiable existence of Spirit and Soul. The study of Energy, and the application of the concepts learned through the study of Energy, is the only true **Operative form of Spirituality**. If it is in fact true that the only constant in the Universe is "Change" and we comprehend that "Change" is caused by a "Force", then we comprehend that

## R.E. ENGINEERING

the Fundamental Forces in Nature are the Ultimate Reality and only thing Consistent in Existence. Because Forces are the CAUSE of effects, we seldom observe and Experience Forces or Energy directly, but rather just observe and experience the EFFECTS of the forces or energy. The fact that the Ultimate Reality of Nature and Existence cannot always be directly observed and experienced is what led to the Religious concept of worshipping an unseen, invisible, hidden (**AMUN**) God. However, one instance when we do get to observe and experience one of the Fundamental Forces in Nature is when we see "**Light**" or what is Scientifically called **Electromagnetic Radiation**. And thus, with the advent of Religion, "Light" became associated with a manifestation of the unseen God.

In Arabic, the word for unseen, invisible, or hidden is **Jinn** or **Ginn** which is also related to the words **Genie**, **Genius**, **Gene**, **Engineer**, and **Gannah** or **Jannah** – the enclosed or hidden garden of paradise in the Islamic religion. In the Islamic tradition, the Jinn are a group of **Etheric beings** similar to the **Angels**, however the Jinn are made from **Smokeless Fire** and the Angels are composed of **Light** (electromagnetic radiation). Consider that in the three main monotheistic religions, there are **7 Archangels** (Micheal, Gabriel, Raphael, Azrael, Uriel, Raguel, and Remiel), and angels are supposedly made of light, and also there are **7-Colors** (Red, Orange, Yellow, Green, Blue, Indigo, Violet) that make up the visible spectrum of Light. The Angels are said to live in the **upper atmosphere** whereas the Jinn live closer to Earth. **Jinn** also have **free-will** like humans and can **reproduce** (**generate**), whereas Angels do not. The word "**Seraph**" (Strong's H8313) or "**Seraphim**" (Strong's H8314) means "**to burn**", "**burning ones**", or "**fiery serpents**", and is the name for an Angelic, Spirit, or Ethereal beings in the Judeo-Christian Bible (Numbers 21:6). Considering that the word "Angel" actually means "messenger", then we can see that the Angels made of Light from religious tradition are actually Electromagnetic Messages much like modern Radio and Television signals. The word **Jinn** or **Ginn** is also related to the word **Genie** (Etheric beings that grant wishes), **Genius**, **Ingenuity**, **Engineer**, and **Genetics** which are all associated with **mental abilities** and **life**. Therefore, there is a literal and etymological connection between the Etheric "Ginn" and the mental "Genius" or "Engineer" and the "Genetics" or "Genus" of life. Even in the more recent Theosophical Philosophies, the Etheric word "**Akashic Records**" is used to refer to a mental Ethereal library of **all knowledge**. One of the 99 names of Allah in Islamic tradition, **Al-Aziz** "the Mighty", actually comes from a pre-Islamic warrior Goddess named **Al-Uzza**. Al-Uzza represented **the morning and evening star** and was worshipped by the **Nabataeans** who associated Al-Uzza with the planet **Venus** and **Lucifer** "the **light bearer** and **morning star**". **Abdul-Uzza** was also the name of the uncle of the Islamic Prophet Muhammad who received the name **Abu Lahab** meaning "**Father of Flames**". However, the name of the **Angel Uriel** means "**Fire of God**" or "**God is my Light**" because Fire is a Natural source of light. Al-Uzza has also been associated with the angel **Metatron** in Judeo-Christian theology.

Ether is needed for life in the form of the air we breathe. Also, our very consciousness and thoughts are made up of Electromagnetic Radiation Ether. In addition to the breath of life and the material of our thoughts, Ether is the "spiritual" realm from religious traditions wherein spiritual or "Etheric" beings such as Jinn, Angels, Spirits, Souls, and God dwell.

**7) Question: How can Æther can Incinerate and Burn?**

**Answer:** In modern chemistry, the word "**Ether**" is used to describe a highly **flammable liquid** that is used as fuel called

## R.E. ENGINEERING

Diethyl Ether. In biochemistry, Ether bonds are found in **carbohydrates**. Diethyl Ether and other flammable liquid mixtures of **hydrocarbons** are collectively called **Naphtha**. The word Naphtha shares the same etymological root as the words Nephesh, Naphash, Nappahu, and Nefu mentioned earlier. As mentioned earlier, Ether is related to the state of matter called Plasma, and Flames are a state of matter classified as Plasma, thus the Flames of a Fire are also related to Ether. The presence of the Energy Element **Æther** in any of the states of matter Solid, Liquid, or Gas has the ability to create Flammable Solids, Flammable Liquids, and Flammable Gases.


The ability of Ether to incinerate and burn is why the Greeks called Africans by the name **"Aithiopian"** literally meaning "Ether People" or **"Burned People"** referring to the dark color of the **melanin** skin pigment. Africans or "Ether-opians" are definitely a **"People of the Sun"** in that the source of the Dark tones of Melanin came from being metaphorically **"burned"** by the Sun. It is a known fact that a Sun tan can be achieved by exposure to **Ultraviolet Radiation**. **Melanogenesis** is the process by which Melanin in the skin is produced. The processes of **suntans**, **sunburn**, and melanogenesis are virtually identical. However, suntans and sunburns do not affect a person's DNA and are caused by frequencies of Ultraviolet Radiation called **UVA** between the range of **750 THz to 940 THz**. However, Ultraviolet Radiation called **UVB** between the frequencies of **940 THz to 1070 THz** can affect a person's genetic **DNA** and causes **Melonogenesis**. The melanin within the skin of African or "Ether-opian" people is indeed a Genetic DNA trait inherited from the Sun. Since UVB Ultraviolet Rays are the fastest frequency of light that come to Earth from the Sun, it is indeed a fact that **the DNA of African or "Ether-opian" people came from the fastest frequencies of Light from the sun**. So indeed, Ether people have Genes determined by the Heat of the Sun, or "**Sun Heat Genes**". Dark colors absorb more frequencies of visible Light, thus the Melanin that exists in African people with Dark skin facilitates the absorption of light energy (Ether) from the sun just like the **Chlorophyll** pigment in plants enables the **Photosynthesis** of Sun energy, and dark **Solar Panels** enable the production of **Solar Energy**. Moreover, **Melanin** is an **electrically conductive** substance used to create **solid-state** "**Organic Electronic**" devices. In an experiment conducted in the 1970s, scientists have shown that Melanin can **emit light** in a process called **Electroluminescence**.

### 8) Question: How is Æther is related to Light?

**Answer:** In modern Science, **Light** is correctly identified as **Electromagnetic Radiation**. It has been shown how **Æther** is related to Electromagnetic Radiation, and thus Æther is also related to Light. However, it must be stressed that Electromagnetic Radiation includes both **Light** and **Dark**, **Visible** and **Invisible** frequencies. Thus, **Æther** is related to both Light and Dark; visible and invisible. Moreover, since Ether was said to be the material in which Light propagated and we know that Light propagates in Electromagnetic waves, then we can observe the direct correlation between Æther, Electromagnetic Radiation, and Light.

### 9) Question: How is Æther is associated with Energy?

**Answer:** It was shown previously how **Æther** is related to the **Fundamental Forces in Nature** namely **Electromagnetic Radiation**, the **Strong Force**, the **Weak Force**, and **Gravity**. **Energy** is the amount of **Work** that can be done by a **Force**, thus Energy is directly related to Forces and **Æther**. The Universe is composed of about 70% Energy Ether. The state of

## R.E. ENGINEERING

matter called Plasma is like a transition between Energy and Matter. After Energy, Ether in the form of Plasma is the most common state of matter in the Universe.

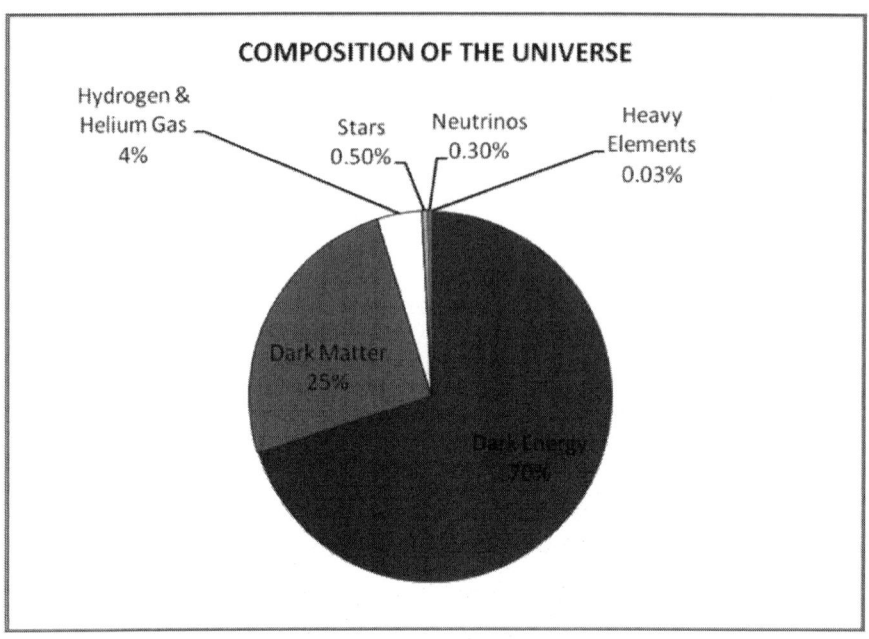

| COMPOSITION OF THE UNIVERSE |||
|---|---|---|
| **Æther** | Dark Energy | 70% |
| | Dark Matter | 25% |
| | Hydrogen & Helium Gas | 4% |
| | Stars | 0.50% |
| | Neutrinos | 0.30% |
| **MATTER** | Heavy Elements | 0.03% |

We experience Ether as **Terrestrial Plasma** here on Earth in the form of **Lightning, Ball Lightning, Saint Elmo's fire, Upper-Atmospheric Lightning,** the **Ionosphere, Polar Aurora, Fire Flames,** and Human-made **Electronically Produced Plasma**. We observe Ether as **Extra-Terrestrial** or **Cosmic Plasma** in outer-space as the **Sun and Stars, Solar Winds, Interplanetary Mediums, Interstellar Mediums,**

## 9 E.T.H.E.R.

**Intergalactic Mediums**, **Flux Tubes**, **Accretion Discs**, and **Interstellar Nebula**. So indeed, <u>the Light or Ether that travels to the planet Earth from the Sun and Stars is in fact empirical evidence of an "Extra-Terrestrial" being that has traveled to Earth</u>. Ether in the form of Plasma has also been suggested by people who study the **Paranormal** as the "substance" which makes up **Ghosts**, **Spirits**, and **Specters**. The name given to the Ether Plasma that is the composition of Ghosts, Spirits, and Specters is **"Ecto-plasma"** meaning "**without form**".

---

The preceding "Questions and Answers" have been presented to illustrate how the **9** Attributes of Æther from Ancient Alchemical Theology correspond to **9** Concepts related to Æther in modern Science. The **9** modern Scientific Aspects of Æther that will be discussed in the following chapters of this book are:

### 9 Modern Scientific Concepts Related to Æther:

1) Antimatter

2) Electrons and Photons

3) Plasma and Ionization

4) Hydrodynamics

5) Electricity

6) Magnetism

7) Electromagnetic Radiation

8) Thermodynamics

9) Resonant Energy

The significance of the number **9** as it relates to concepts associated with Ether, coupled with information presented by Dr. Malachi York about **"9 Ether"** is why this book has been entitled **"9 E.T.H.E.R. R.E. Engineering"**.

www.AfricanCreationEnergy.com

# R.E. ENGINEERING

**Question: What does Dr. Malachi York Teach about 9 Ether?**

**Answer:** Dr. Malachi York has discussed 9 Ether in particular and Ether in general in various books that he has authored over the years:

"**Nine Ether** is the Original Creative Forces that made these booklets possible. **Ether** means in Middle English 'Upper Air' From Latin Aether and Greek Aither. Ghost is the Lowest and Ether is the Highest **Anu** or **On**. The Forces of **9-Ether**, called Black Forces created Life in the water first. Hydrocarbons were present in these waters. Hydrocarbon is a simple **Methane** or **Benzene** Gas, which only contains Hydrogen and Carbon. The word Hydrocarbon itself means **Hydro- Hudor** 'We or **Water**' and Carbon-Carbo 'Charcoal Black'. So these Black Gases of **9-Ether** are the Celestial Origin of all Nuwaupians..."

~The Sacred Records of Atum-Re "The Black Book part 2" Chapter 10, Scroll 10, page 185

"Nine ether is the combination of all existing gases of nature. Nothing anywhere can be as powerful as all the existing gases. On earth these gases are known as, **Radon (Rn)** with an atomic number of **26**, **Xenon (Xe)**, with an atomic number of **54**, **Krypton (Kr)**, with an atomic number of **36**, **Argon (Ar)**, with an atomic number of **18**, **Neon (Ne)** with an atomic number of **10**, and **Helium (He)** with an atomic number of **2**. These are also called **The Noble Gases** on a periodic or elemental chart, on the physical chart. However on the ethereal chart, they are listed as thus, $E_2$, $E_{10}$, $E_{18}$, $E_{26}$, $E_{36}$, $E_{54}$, and not the word element, elementary, from elementum, 'first principle, rudiment, beginning.' Used as elementary, the beginning without importance, as of yet. Therefore, 9 ether is the most potent power in all the boundless universe."

~ The Holy Tablets, Chapter One: The Creation, Tablet One: Epic of Creation and Before, page 3

# 9 E.T.H.E.R.

*"Physical beings vibrate on the physical plane called nassuwt. Naasuwt is 1-solid, 2-liquid, 3-gas. You as a physical being have a spirit in it. The spirit in you descended down from etheric, the physical you descended down into a physical body...The etheric beings travel down and materialize into the body. The 'supreme beings' are living in the realm of malakuwt as it is called in the Galilean (Arabic) language, which is the 4th plane, malakuwt. Malakuwt is a pathway that spiritual beings come into. Spiritual beings can travel. Extraterrestrials vibrate on this realm and start going up higher. When spiritual beings reach the 4th realm, then they'll decide to come to the physical plane."*

~**What is Spirit and Soul page 50**

| ABODE | PLANE | ETHER | FUNCTION |
|---|---|---|---|
| LAAHUWT | 7 | 4TH Ether: | Mental Reflecting |
|  | 6 | 3rd Ether: | Light Ether |
| MALAKUWT | 5 | 2nd Ether: | Life Ether |
|  | 4 | 1st Ether: | Electric Ether |
| NAASUWT | 3 | Gases: | Dense Physical |
|  | 2 | Liquids (Chemicals) |  |
|  | 1 | Solids Region |  |

"The ultra gaseous matter, 'ether' in the physical form is associated with the solids of the body such as the structure of the body. The life force 'ether' which is etheric form is involved in the fluids of the body called liquid ether such as the blood stream and the excretions of the ovaries and testes. The spirit is in the blood. The blood is liquid. That <u>spirit acts a spark that ignites</u> the life in the blood. The life produces a light, which appears as an <u>auratic shell</u> it in-houses the soul which is the light messengers. This messenger transports information to the conscious and sub-conscious being. The sparks of life trigger the mind that pulls itself from the mental reservoir of intellect, which is governed by the oversoul that is connected by an etheric cord. All messages in light forms travel through the cords from the mental reservoir of intellect called Akasha. Each being receives <u>light messages</u> in their DNAs and RNAs from and through these beings that are responsible for triggering each moment of thought and action in their existence.

# R.E. ENGINEERING

*So, a harmonious flow of energy and life in the form of light messages must travel to the being rendering each being's inner self etheric."*

~ **The Holy Tablets, Chapter Seven: The Living Soul, Tablet Four: ETHER, page 702**

"Nine ether will become 6 ether through time and age (die). 6 ether becomes ghost through time and age. Six ether is 9 ether in death, and ghost is the death of 6 ether. After the death of 6 ether, 9 ether resurrects again."

~ **The Holy Tablets, Chapter One: The Creation, Tablet One: Epic of Creation and Before, verse 24-26**

"Etherians are 720 degrees (360 Square and 360 Circle)...The Etherians Are Crystal Light Energy From ILLYUWN (the **Highest Place**) And Can Take On The Form Of Anything...The Etherians Are The **Providers** (RIZQIYIAN) And They Police The Universe In Respect, As Far As Being Responsible For Living Things. We Like Good Angels Because They Do Things For Us; And We Don't Like Bad Angels Because They Hurt Us. They Are Not Concerned With This, But Must Maintain The Universe To Prevent A Star Holocaust. They Personify To Whomever And Their Assignment Is The Guardians Of Life... They [Rizqiyian] Perspire And Transform From Rizqiyian To Etherian, And From Etherian To Rizqiyian. You Transform Too. Every Time You Perspire Something Leaves From Your Body...THE RIZQIYIAN Claim To Be The Parents Of The Nubians, They Are The Builders Of The Pyramids...The RIZQIYIAN Are Physical Beings Evoluting Towards Etherians, Energy Or Light Beings. The Etherians, Who Are The Guardians, Are Equivalent To What You Call Spiritual Or Angelic Beings. Their Main Concern Is The Intellect..."

~ **The Man from Planet Rizq, page 109-111**

## 9 E.T.H.E.R.

**Question: Are there other writers who Teach about 9 Ether?**

**Answer:** Yes, 9 Ether is discussed in books called "The Nine Ball Liberation Information counts I-IV" by the authors Wu-Nupu, Asu-Nupu, and Naba Nupu:

"NINE ETHER is the <u>mental power</u> and chemical forces that GREW the Universes. <u>Nine Ether is THE COMBINATION OF ALL EXISTING GASES and CHEMICALS</u>. Nine Ether is the ORIGINAL FORCES OF ORDER...Nine Ether began as a SPONTANEOUS FIRE in Primeval Chaos and grew bodies for itself called TRUE STARS or SUNS, one of which is THE SUN of our Solar System...<u>Nine Ether produced the bodies of the Suns and in turn the Suns produced Nine Ether</u>...Nine Ether is BLACK, therefore, the bodies of the Suns are Black (JET BLACK). Nine Ether is our powers in Nature and by Nature."

~The Nine Ball Count I Liberation Information page 23

"SIX ETHER is Nine Ether IN DEATH. To put it another way, when Nine Ether dies, it becomes Six Ether. As stated in COUNT I of the booklets THE NINE BALL, Nine Ether is the completed combination of all existing gases and chemicals, and there can be nothing more powerful – this is why Nine Ether is the Original Creator or GROWER. In Ethiopian Culture, CREATION means GROWTH. Therefore, Six Ether is the reverse formula of Nine Ether. Nine Ether is the formula of life-and Six Ether is the formula of death, and this means that the original life-giving gases and chemicals in Nine Ether dissipate and lessens to the point of ineffectiveness and this marks the death of Nine Ether"

~The Nine Ball Count II Liberation Information page 23

# R.E. ENGINEERING

**Question: Are there any other sources who Teach or discuss 9 Ether in particular and Ether in General?**

**Answer:** Yes, the writer named "Afroo Oonoo" discusses 9 Ether, Natural Ether, Universe Ether, and Six Ether in the book "Introduction to the Nature of Nature":

"**NATURAL ETHER** (Universe Ether) is THE BLOOD, MIND, and ENERGY of The Universes. Natural Ether is the Energy that gives **light** and **life** to the Universes and the person and things therein...UNIVERSE ETHER is the **combination of all elements** and all kinds of Matter that can become energy and have become energy to the degree that can exist at any given point on The SMAT Circle of Order...Helper-Ether (Sunshine) is the physical life and light of the World, and Head-Ether (Reason) is the mental and spiritual light and life of the World...NATURAL ETHER IS ALL ETHER produced by NATURAL PROCESS, and UNIVERSE ETHER is ALL ETHER produced by UNIVERSE ORBS and is likewise by natural process...Natural Ether is **Starshine (Sunshine)** and THE **HEAT** of the World...Basically and scientifically, there are two major kinds of **Natural Ether (Universe Ether)**, namely: **NINE-ETHER** and **SIX-ETHER**. Natural Ether is **NATURAL ELECTRICITY** and **ETHEREAL ELECTRICITY** which is life, light, and mind energy...Natural Ether (Universe Ether) is THE COMBINATION of, all different kinds of elements of Matter and all types of Matter that can become active energy and have become active energy like creative fire, active blood, and active mind... Nine Ether (The Positive) controls and supports The Living, and Six Ether (the Negative) supports and controls The Dead...Nine-Ether is THE LIFE of THE SCIENTIFICALLY LIVING and Six-Ether is THE LIFE AND DEATH OF THE SCIENTIFICALLY DEAD...Nine-Ether (The Positive) is called **NOOPOOH** and Six-Ether (The Negative) is called ZOOPOOH...ORIGINAL NATURAL ETHER, like all Universe Ether that follows, is produced by The Activating and Motivating Suns of The Elements, and is The **Active Ether** (The **Creative Fire**) that put (grew) The Universes into creation order...

## 9 E.T.H.E.R.

*THE LAW OF NATURAL ETHER* states: "The **Energy of Energies** is life or death called NATURAL ETHER (Universe Ether) who is all that energy can be at any given time and does all that energy can do at any given time while being the results of Brain-Noots melting down by way of Brains, the same way each time that conditions are the same, and in accord with The Nature of Nature. In Nature Science, active energy is active Matter and active power. The Ethiopian Race got its name from the, root-word ETHER which means BURNING LIKE LIFE and BLACK LIKE VACUUM...Physical and visible **light**, Sunshine, The Suns themselves, **fire**, **heat**, **electricity**, Natural-Universe Ether, and even life itself (because life is a burning) are BLACK LIKE VACUUM, and this is why these energies turn persons and things BLACK whenever they come in contact with these powers long enough... Natural Ether is **Creative Fire** (Fire of Growth) or **Destructive Fire** (Fire of Decay). It creates life and creates death. Ether creates LIFE OF LIFE by origin and development called growth or increase in chemical strength by organization, and creates LIFE OF DEATH by decay or decrease in chemical strength by deterioration and disintegration...HEAD-ETHER, also called **CREATIVE ETHER** and GUIDING ETHER, is THE ENERGY REASON who is THE COSMIC and MUNDANE INTELLIGENCE."

**~Introduction to the Nature of Nature Book 1, page 59-64**

# R.E. ENGINEERING

"Listen to Reason! When The Natural Electricity (Ether who is The Burning) reaches the brain, it lights up THE BRAIN like a bulb, then the person not only has LIFE but CONSCIOUSNESS with that life, because he or she can then see, feel, taste, hear, and smell. Therefore, a person sees, feels, tastes, hears, and smells with his or her mental organ (the Brain), the organ of consciousness, by way of the nervous systems of the brain."

~**Moonset and Sunrise in the Nature of Nature, page 29, paragraph 37**

"Listen to Reason! Like the engine of a car gives off (exhausts) CARBON MONOXIDE, the lungs of a person give off CARBON DIOXIDE. Carbon is residue of a BURNT substance or gas. The point is this: An individual inhaling OXYGEN and exhaling CARBON DIOXIDE is EVIDENCE of BURNING taking place in the lungs and throughout the circulatory system of a living person, and this is why the body of a living person is WARM and that of a dead person is COLD. Open your mouth and blow your breath on the back of your hand and feel how warm it is, and you will know that burning is taking place --oxygen being burnt by blood and creating **NATURAL ELECTRICITY** called **NATURAL ETHER** which is LIFE. Then too, what about that STATIC ELECTRICITY that causes clothes to cling to body --this is further proof that life is indeed NATURAL ELECTRICITY known as NATURAL ETHER who is the life of all living persons and things in one degree or another throughout The Universes. Let it be remembered always and let it be known!"

~ **Moonset and Sunrise in the Nature of Nature, page 30, paragraph 38**

## 9 E.T.H.E.R.

A "Re-introduction" is a proposal of something previously rejected, and since the concept of Ether was initially accepted, but then rejected by the scientific community, this book entitled "**9 E.T.H.E.R. R.E. Engineering**" is a "Reintroduction" to the concept of Ether which does have operative, practical, and applicable scientific implications. This book also provides several "**Ether Experiments**" to provide the reader with first-hand Experience and Evidence of Ether in Application. This book is written for the **Ether-opians (Ether People)** who study information about Ether to become **Ether-ologist**, and apply the information to become **Ether Engineers** and **Ether Technicians**. We affectionately call practitioners of Ether knowledge by the name "**Ether Benders**". There are **9** Areas where the Application of Ether knowledge can be utilized:

1) **Stoichiometrist** - scientist of explosions and chemical reactions

2) **Pyrotechnicians** - technicians of explosions

3) **Firefighters** - managers and controllers of fires, flames, and life

4) **Electrical Engineers** - scientist of Electricity

5) **Electricians** - technicians of Electricity

6) **Spectroscopist** - light and LASER scientist

7) **LASERist** - light and LASER technicians

8) **Light Workers** – Spiritual/Esoteric study and application of Ether

9) **Ether Benders** – Masters of all fields of Ether Application

The first letter in the names of chapters 2 through 7 of this book are the letters for the word AETHER (Antimatter, Electricity, Thermodynamics, Hydrodynamics, Electromagnetic Radiation, Resonant Energy). This book entitled "**9 E.T.H.E.R. R.E. Engineering**" is a continuation of the work of "African

## R.E. ENGINEERING

Creation Energy" to transform "Speculative African Theologies into operative African Technologies". This book builds on the foundation laid by **"African Creation Energy"** in the book entitled **"P.T.A.H. TECHNOLOGY: Engineering Applications of African Sciences"**. This book is an operative book written specifically for the purpose of use, practice, and application. This book talks about the African origin of principles related to Science and Engineering as it relates to Ether as it was known in Ancient African Alchemy. This book also provides Experiments to practice, apply, and experience the principles presented within the book.

**Technology** is the operative manifestation of **spirituality**. Thus, you know if your spiritual science is Right and working when it manifests into the real world in the operative form of Technology (which means to apply knowledge). Knowing that Ether is "Spirit", then Ether Engineering empowers the practitioner with the ability to operatively **control "Spirits"**. As people of the sun, Ethiopians or "Ether Utopians" who descended from Africa, our spiritual science should enable us to manipulate all of the Alchemical elements, thus we are "The **First Ether Benders**", and this book entitled **"9 E.T.H.E.R. R.E. Engineering"** along with "**P.T.A.H. Technology**" provides a scientific framework to achieve that goal. This book entitled **"9 E.T.H.E.R. R.E. Engineering"** teaches the reader how to literally and metaphorically "Harness the Power of Fire", or **"Harness the Power of Ether"**.

## 2.0. ANTIMATTER

One of the attributes of Ether is that Ether is Primordial, Essential, and Fundamental. In modern Science, the Fundamental or Elementary Particles of Nature are the smallest building blocks that make up all matter in existence. However, there are also Fundamental Particles that make up **Non-existence**, and these particles are called **Anti-Particles**. Anti-particles have the **opposite Electrical charge** of the Fundamental Particles of Nature. When an Anti-particle encounters its opposite Fundamental Particle, the Anti-particle and Fundamental Particle **Annihilate** each other and become **Energy particles**. The word "Annihilate" comes from the Latin word *nihil* and means "**to make into nothing**". Anti-particles can combine to form **Antimatter**. If we consider Matter as the material of the "Material World" (Something-ness), then we can consider Antimatter the material or "stuff" that makes up the **Non-existent** or "**Immaterial world**" (Nothingness). It stands to Reason that "Nothing" or a "state of Nothingness" would have to exist prior to the creation of "Something". Thus, it is possible that the Anti-Particles of Antimatter are indeed the **Primordial "Abyss"** from which the Fundamental Particles of Nature emerged. In Ancient African culture in Egypt, the Principles of the Primordial state at the beginning of creation were represented by 8 deities called the Ogdoad or **Khemenu**. The Khemenu were 4 Males who were depicted with the head of frogs, and 4 Females who were depicted with the heads of snakes. The names of the 8 Khemenu were *Amun* (Hidden), *Amunet* (Mystery), *Kek* (Darkness), *Keket* (Void), *Heh* (Infinity), *Hehet* (Eternity), *Nun* (Primordial Abyss), and *Nunet* (Chaotic Waters). The 8 Khemenu, who could have represented Antimatter, collectively existed in a state called **Nun** or **Nu**.

# R.E. ENGINEERING

The 8 Ogdoad or Khemenu of Ancient Egypt –
Symbolic of Ether in the form of Antimatter

## 9 E.T.H.E.R.

Amongst the **Dogon** tribe in **Mali, West Africa**, the **8** primordial deities of creation are called the **Nommos**. In Ancient Egypt, the **Underworld** or *"the other side"* was called the *Duat* or *Amen-Ta* and was also connected with Nun, the Khemenu, and possibly Antimatter. The relationship between the **8 Khemenu** of Ancient Egypt to Antimatter is even more remarkable when we consider that there have only been **8 Anti-particles** that have been experienced and scientifically proven to exist to date.

| # | Antimatter "ETHER" | Matter Element |
|---|---|---|
| 1 | Photon | Photon |
| 2 | Anti-Neutrino | Neutrino |
| 3 | Positron | Electron |
| 4 | Anti-Quark | Quark |
| 5 | Anti-Neutron | Neutron |
| 6 | Anti-Proton | Proton |
| 7 | Anti-Hydrogen | Hydrogen |
| 8 | Anti-Helium | Helium |

**8 OBSERVED ANTI-MATTER PARTICLES**

The first Anti-particle that was observed experimentally was the Positron in 1929. The Antiproton was observed in 1955 and the Antineutron was observed in 1956. Anti-quarks and Anti-Helium were both observed in the 1970s. The most recent Anti-particle to be observed was Anti-Hydrogen in 1995. The theory of Anti-particles and Antimatter suggests that Photons and Neutrinos are their own Antiparticles. Antimatter theory further suggests that Anti-particles can be combined to form Antimatter elements much like Fundamental Particles combine to form the elements of the Periodic Table. Thus, theoretically, an "Antimatter Table of Ethers" can be constructed on the "other side" of Hydrogen much like the Periodic Table of Elements has been created on "this side of Hydrogen".

# R.E. ENGINEERING

The image below shows the **Periodic Table of Elements** (right) with their reflections in The **Anti-Matter Table of Ethers** (left)

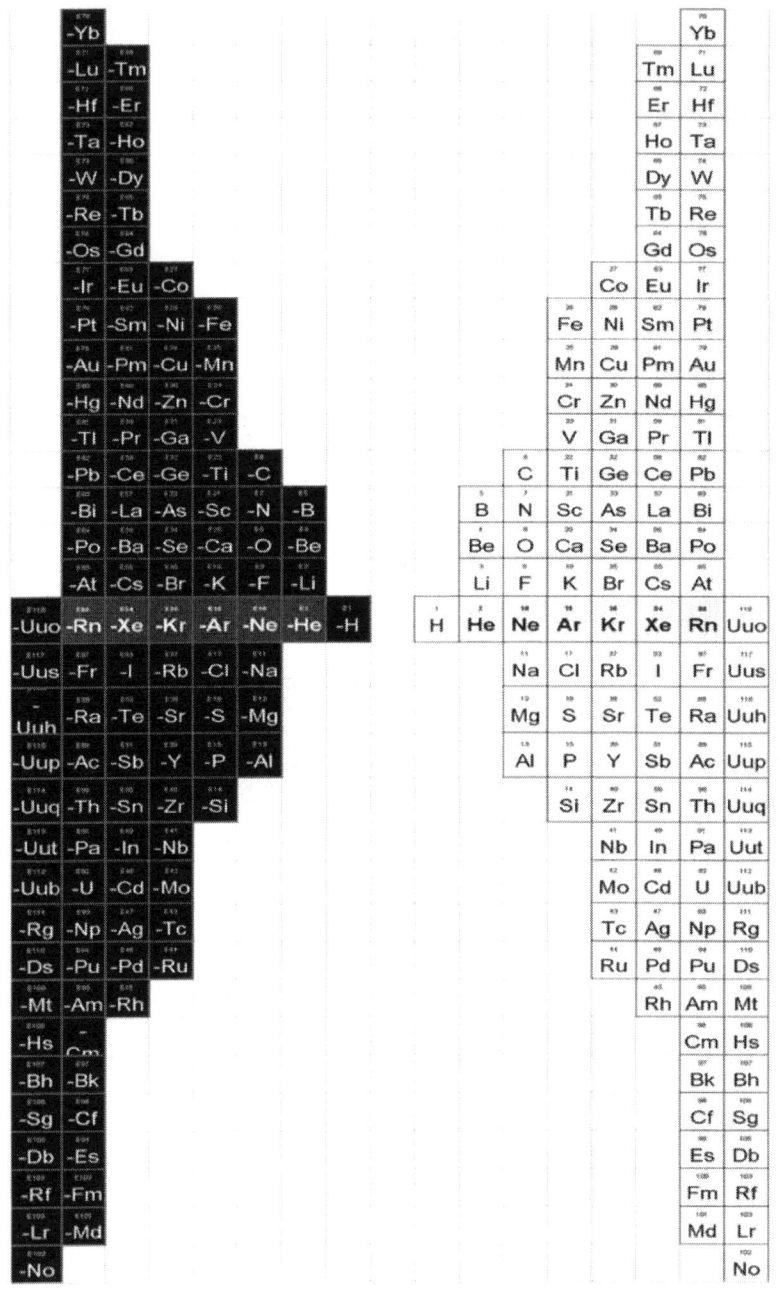

www.AfricanCreationEnergy.com

## 9 E.T.H.E.R.

Whereas Helium, Hydrogen, Protons, Neutrons, Quarks, and Electrons all have observed reflections in Anti-matter as Anti-Helium, Anti-Hydrogen, Antiprotons, Antineutrons, Antiquarks, and Positrons (Anti-electrons) respectively, Photons and Neutrinos are interesting because they are their own anti-particle. **Photons** and **Neutrinos** being their own Antiparticles is similar to **Amun** and **Amunet** of the Ogdoad Khemenu who are the only 2 of the 8 Ogdoad who exist both in and out of the state of *Nun* or *Duat*. Photons are energy particles of Light or Electromagnetic Radiation denoted by the Greek letter Gamma "$\gamma$". Neutrinos are subatomic particles denoted by the Greek letter **Nu** "$\nu$", that are able to pass through matter without being affected or detected. Since billions of Neutrinos pass though your body every second without being detected, Neutrinos are considered **omnipresent** in Nature. Photons and Neutrinos are created by atomic radioactive decay and also Annihilation between matter and antimatter. While current Physics theories suggest that nothing in the Universe can travel faster than the speed of Light (Photons), several experiments in September and October of 2011 as well as in prior years have shown that **Neutrinos may have the ability to travel faster than Light** (Photons). The main reason why Neutrinos may have the ability to travel faster than light is because Neutrinos are not affected by matter, and therefore would not slow down the same way Photons slow down when traveling through matter. Photons are particles of light (electromagnetic radiation), and since Neutrinos are undetectable by our Human senses, Neutrinos can be considered **"particles of dark"**. If these dark particle Neutrinos can indeed travel faster than light, then they are empirical evidence of the hypothetical "faster than light" **Tachyons**. Proponents of Tachyon theories suggest that these faster-than-light particles could facilitate **Time Travel**.

## 3.0. ELECTRICITY

Comprehension of the **Electron** and **Electricity** is prerequisite to the discussion of the relationship between **Ether** to Electromagnetic Radiation (Light), Plasma, and Ionization. At the time of this writing, technology has not advanced to the point where most people will be able to have a first-hand experience with Ether in the form of Antimatter. However, after Antimatter, the smallest fundamental particle that most people are able to have experience with is the Electron and Electricity. **Electrons are the workmen in nature**. Electrons are constantly working or moving, orbiting the nucleus of atoms. Also, whenever there is movement or animation at any scale or size, there are always electrons moving. The flow or movement of Electrons is called "Electricity". The book **"P.T.A.H. Technology"** by "African Creation Energy" explains how the cosmology and philosophy associated with the African Creation Deity of the **workmen** named **PTAH** is related to "**Ohm's Law**" and **Electric circuit theory**.

Above: African Creation Deity PTAH symbolic of Electric Ether Theory

## 9 E.T.H.E.R.

## PTAH'S LAW OF ELECTRONICS

POWER
VOLTAGE
CURRENT
RESISTANCE

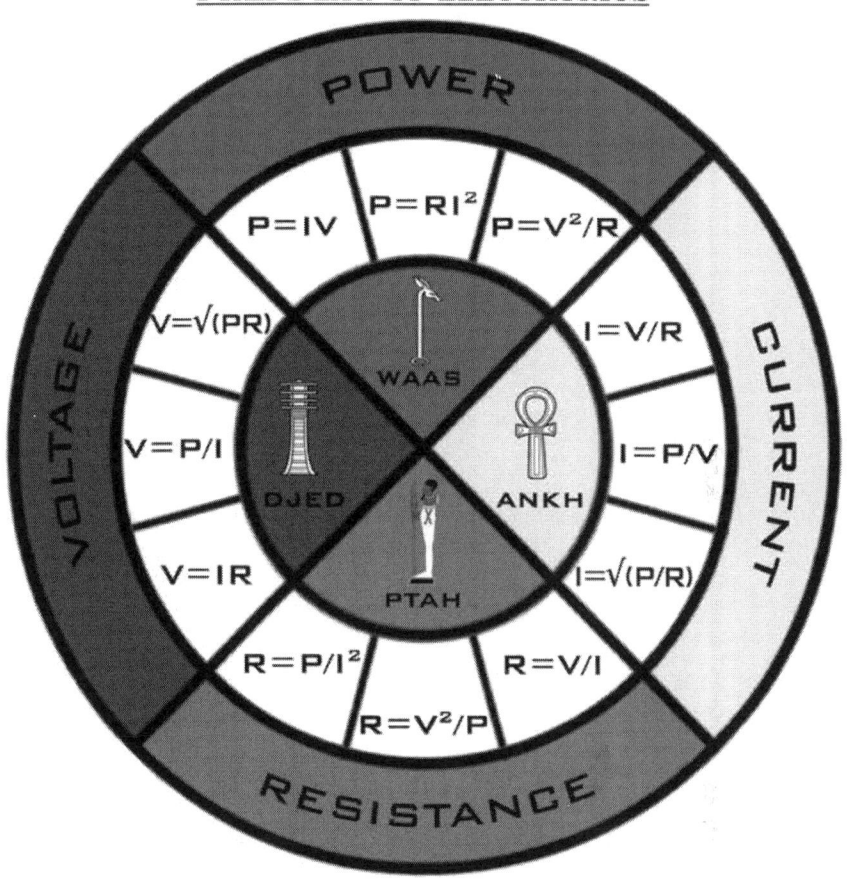

$P=IV$
$P=RI^2$
$P=V^2/R$
$V=\sqrt{PR}$
$I=V/R$
WAAS
$V=P/I$
DJED
$I=P/V$
ANKH
$V=IR$
PTAH
$I=\sqrt{P/R}$
$R=P/I^2$
$R=V/I$
$R=V^2/P$

"Ptah's Law of Electronics" showing modern mathematic equations for the Electronic relationships and symbolism of the components that combine to form "staff of Ptah" circuit.

Djed, Ankh, Waas, and Ptah are represented with the modern mathematical symbols V, I, P, and R representing Voltage, Current, Power, and Resistance measured in Volts, Amperes, Watts, and Ohms respectively.

Just as the **electron** is the smallest fundamental particle to **rise from the primordial abyss of Antimatter**, it is said in Ancient African cosmology that **PTAH rose from the primordial abyss of Nun**. Also, just as the movement or **vibration of electrons can cause light**, it is said in Ancient African cosmology that the "speech" or **vibration from PTAH** gave birth to **Atum** or Atum-**RE** who is symbolic of **light**. In the Memphite Theology, PTAH creates by way of symbolic contemplation and speech, thus PTAH is also representative of the mental faculties of the mind, consciousness, and Reason. The relationship between thoughts, reason, electrons, and electricity is that thoughts and reason are actually electrons flowing within the brain. The very substance of your thoughts is electrons. A thought is an electrical signal sent between neurons in your brain. Thus, both **conscience** and **conscious** are composed Ether in the form of flowing electrons. It we consider that Plasma is an Ionized Gas, and an Ionized Gas facilitates the flow of electrons, and conscience and conscious are composed of flowing electrons, then it is indeed possible that Ionized Gas in the form of Plasma (Ether) could be both **conscience** and **conscious**. Comprehending the fact that thoughts and reason are composed of the fundamental particle called the electron, then we must also comprehend that relative to the **Ether of the mind**, there are **3 creations**: **1) Quantum-sphere** – the realm of **electrons**, **2) Noosphere** – the realm of **human and other biological thoughts** within brains in general created by electrons, and **3) Cyber-sphere** – the realm of **computers and mechanical thoughts** within machines in general created by brains which were created by electrons. Therefore, it is no surprise that the ability of computers to speak or network with one another through the flow of electrons and electromagnetic radiation has been termed an "**Ether Network**" or **Ethernet**.

## 4.0. THERMODYNAMICS

One of the aspects of **Ether** is that Ether is related to **Burning**. In modern science, **Thermodynamics** is the study of **Heat, Energy, Temperature, Pressure**, and **Work**. Mastering the Science of Thermodynamics enables the practitioner to have complete comprehension of *"life and death"*, **energy transformation**, and the processes in Creation in general. The etymology of the word "Thermodynamic" comes from the Greek words *"Thermos-"* meaning "Heat" and *"-dynamics"* meaning "Power". Thus, the word "Thermodynamics" literally refers to the **"Power of Heat"**. In Ancient African cosmologies, the Power of Heat was personified as a Goddess named **Sekhmet** whose name literally meant **"the Powerful one"**. Sekhmet was the **daughter of the sun** god RE who was said to have created her from his *fiery eye*. Sekhmet was said to represent the **Heat of the Sun at Noon** and she was given the title "Nesert" or **"Lady of Flame"**. Sekhmet was the **wife of Ptah** and was the moderator of **Ma'at** or **Equilibrium**. Sekhmet was seen as **both a creative and destructive force**. Because of the potentially destructive qualities of Sekhmet, she was also seen as a Goddess of War. The "priests" and "priestesses" of Sekhmet were doctors, physicians, and healers. Sekhmet was depicted as a lioness or a woman wearing a feline lioness mask. Amongst the Southern Egyptians, Kushites, Nubians, and Central Africans, Sekhmet was synonymous to the name **NYABINGHI** (also spelled Nyabinghi, Nyahbinghi, or Niyahbinghi). The oral tradition states that Queen Nyabinghi was Kushite princess who rebelled against injustice in her kingdom as a child and fled to the Congo forest where she organized a guerilla army and retook her throne becoming the **Queen of Queens**. Various female freedom fighters over the years have been said to be possessed by the "spirit of Nyabinghi". The word "Niyabinghi" means "Black Victory" coming from the words "**Niya-**" meaning "Black" and "**-binghi**" meaning "Victory". Niyabinghi is an important principle amongst the **Rastafari** movement (*Pure Fire - Purifier*).

**Heat** is scientifically defined as the **Transfer of Energy** from one thing to another by way of contact or radiation. Heat is the fundamental process of **Energy Transfer** between objects, and **Thermal Radiation** is one of the fundamental methods of Heat Transfer between objects. One of the best known sources of Heat is Fire. **Fire** is scientifically defined as the rapid "**Oxidation**" or **loss of the electrons** of a substance in a **chemical reaction** that gives off **Heat** and **Light** in the form of **Thermal Radiation**. In contrast to Oxidation, **Reduction** is the **gain of electrons** by a molecule, atom, or ion. An **Anti-Oxidant** is a molecule capable of inhibiting oxidation.

Queen Nyabinghi - the feline Lioness Goddess SEKHMET Ancient African Egyptian personification of "the Power of Heat" (Thermodynamics).

# 9 E.T.H.E.R.

The light-emitting gas component of a fire is the Flame. The Flames of hot fires can become an ionized gas or Plasma. **Flammability** is the measure of a materials ability to burn, ignite, or catch fire. There are materials in all states of matter that are nonflammable and flammable: **Flammable Solids**, **Flammable Liquids**, and **Flammable Gases**. **Dimethyl Ether** and **Diethyl Ether** are two examples of flammable liquid (**Fire Water**) mixtures of **hydrocarbons** called **Naphtha**. The color of a Flame depends on what is being burnt. The table below shows various elements that give off Flames with colors of the visible light spectrum when burned.

| Flame Color is Influenced By the Elements Being Burned | |
|---|---|
| **Flame Color** | **Burned Element** |
| RED FLAME | Lithium, Strontium |
| ORANGE FLAME | Calcium |
| YELLOW FLAME | Sodium |
| GREEN FLAME | Boron, Copper |
| BLUE FLAME | Copper, Cesium |
| INDIGO FLAME | Cesium |
| VIOLET FLAME | Potassium |

As we will discuss in the section on Electromagnetic-Radiation, all of the colors of the Visible Light spectrum combine to form White Light. The hottest Flame color is White. Blue Flames are hotter than Red, Orange, or Yellow Flames. Theoretically, **Black Flames** which emit a **"Black Light"** are the coolest.

## R.E. ENGINEERING

Hydrocarbons can form mixtures of Flammable Gases that can emit **Heat** and **Light** when **ignited** and **burned**. Recall that the mental faculties of the mind are composed of electrons flowing through the brain. Thus **"Consciousness" is made possible by the flow of electricity through a substance**. If an **electric current** is passed through a Gas (making it a **"conscious gas"**), then the electrons within the gas will become excited and release energy in the form of **Light**.
**Thermocouples** are electronic devices that can **transform Heat into Electricity**. For the 6 Noble Gases, each gas emits a different color of the Electromagnetic Spectrum when they become "conscious" (that is to say, when an electric current passes through the Noble Gas).

| Color of Light Emitted by Electrified Noble Gases ||||
|---|---|---|---|
| Color Emitted | Symbol | # | Noble Gas |
| Red | He | 02 | Helium |
| Orange | Ne | 10 | Neon |
| Yellow | Ar | 18 | Argon |
| Green | Kr | 36 | Krypton |
| Blue | Xe | 54 | Xenon |
| Chartreuse | Rn | 86 | Radon |
| * An Electric Current passing through a mixture of all the Noble Gases combined emits a Brown colored Light ||||

An important concept in the field of Thermodynamics is called **Entropy**. The word "Entropy" etymologically means to "turn in" and is used in Science to refer to a "measure of the **disorder**, **randomness**, or **chaos** of a system" or a "measure of the energy not available to do work" within a system. The concepts of Heat, Work, Energy, Entropy, Temperature, and Pressure are established and expressed in four laws which typify **Thermodynamic Systems**.

The **Four Laws of Thermodynamics** are:

**The Zeroth Law** — **TEMPERATURE**: *"If two objects or systems are in Thermal Equilibrium with a third object or system, then they all are in Thermal Equilibrium with each other."*

**The 1st Law** — **CONSERVATION OF ENERGY**: *"The internal energy of a closed system equals the Heat transferred into the system minus the output work done by the system. Energy is not created or destroyed but merely transforms and flows from one place to another."*

**The 2nd Law** — **THERMALIZATION**: *"The Entropy of a system that is not in Thermal Equilibrium will increase until a state of Thermal Equilibrium (maximum entropy) is reached."*

**The 3rd Law** — **RESIDUAL ENTROPY:** *"At a temperature of Absolute Zero, a system will have a minimal amount (typically zero) of entropy"*

The Four Laws of Thermodynamics are useful in comprehending **Thermodynamic Cycles**. Thermodynamic Cycles are cyclic processes which transfer Heat, Energy, and Work between states as temperature and pressure changes. The science of Thermodynamic Cycles is applied to construct **Heat Engines** which use the principles of Thermodynamics to do some form of work or accomplish some task in an environment. The efficiency and various stages of the Thermodynamic Cycles of a **Heat Engine** is described as a **Carnot Cycle**.

# R.E. ENGINEERING

## Thermodynamic Cycles in African Theology and Theory

| Thermodynamic Symbol | African Culture | Ether Name | Theology / Theory |
|---|---|---|---|
|  | Congo | Dikenga | The Four Movements of the Sun: 1-Kala: morning, birth - colored black. 2-Tukula: noon, the prime of life - colored red. 3-Luvemba: sunset, late life, death - colored white. 4-Masoni: midnight, rebirth, resurrection - colored yellow |
|  | Akan | Gye Nyame | *Abode santann yi firi tete; obi nte ase a onim ne ahyease, na obi ntena ase nkosi neawie, GYE NYAME.* This great panorama of creation dates back to time immemorial; no one lives who saw its beginning and no one will live to see its end, except Nyame. |
|  | Tchokwe | Lusona | At the top is "God", left is the Sun, right is the Moon, and bottom is a human. The Sun, the Moon, and Humanity all have thermodynamic cycles established by "God". |
|  | Nuwaupu - Noone | 9 Ether - NoopooH | 9 Ether in Death is 6 Ether. 6 Ether in death is 3 Ether, and eventually 9 Ether Resurrects again. Four cycles: Revolution (3 to 6), Origination (6 to 9), Dorigination (9 to 6), Evolution (6 to 3) |

## 9 E.T.H.E.R.

African cosmologies, theologies, and theories are replete with concepts related to Thermodynamic cycles. Studying the **Sun** as a major source of **Heat** and **Energy** on Earth and studying the daily **cycles of the sun** likely inspired the development of theories related to **Thermodynamic Cycles**. The Ancient Egyptians acknowledged Thermodynamic cycles by personifying "four stages of the sun" as Kheper-Re, Aton-Re, Atum-Re, and Amun-Re. Other examples of Thermodynamic Cycles in African theology are the "Gye Nyame" cosmogram of the Akan in West Africa, the Dikenga cosmogram of the Congo in Central Africa, the Lusona cosmogram of the Tchokwe of South Africa, and the Circle of "9 Ether" of the Nuwapians in the African Diaspora. The similarities between these African concepts to modern scientific concepts related to Thermodynamics are astounding. In fact, the transformation or movement through each point of the Dikenga cosmogram is called "**dingo-y-dingo**" which means "**coming and going from the center**" which is similar in definition to the modern scientific term "**Entropy**" which literally means "**to turn into the center**". Modern scientific concepts related to Thermodynamic Cycles consist of the four Thermodynamic processes of **Heat Addition**, **Expansion**, **Heat Removal**, and **Compression** which can be represented as the cycles of **Birth**, **Life**, **Death**, and **Resurrection** or the seasonal cycles of **Summer**, **Fall**, **Winter**, and **Spring** respectively. In modern science, the most efficient Thermodynamic Cycle for **converting Energy into Work** is called the **Carnot Cycle**. The Carnot cycle is named for a **French Military Engineer** named **Nicolas Léonard Sadi Carnot** who was part of **Napoleon's fleet** during the **conquest of Egypt** between the years 1798 to 1801. In 1824, Carnot provided a successful theory on the Thermodynamic Cycles of Heat Engines in a book entitled "**Reflections on the Motive Power of Fire**". Carnot died at the age of 36 and is considered the "Father of modern Thermodynamics". The table on the next page compares the four Thermodynamic processes of the Carnot cycle to four processes found in African theologies related to Thermodynamics.

# R.E. ENGINEERING

| | Thermodynamic Carnot Cycle | Nuwaupu | Dikenga |
|---|---|---|---|
| EXPANSION | Isothermal Expansion: Constant "Hot" Temperature gas expands through heat absorption, work is done on the surroundings (environment) | 9 Ether Dies becoming 6 Ether | Tukula Falling |
| HEAT REMOVAL | Isentropic Expansion: Constant Entropy, Expansion of gas until cool, output works on surroundings (environment) | 6 Ether Dies becoming 3 Ether | Luvemba Falling |
| COMPRESSION | Isothermal Compression: Constant "Cold" Temperature compression of the gas through heat loss, the surroundings (environment) begin working on the gas | 3 Ether resurrects to 6 Ether | Masoni Rising |
| HEAT ADDITION | Isentropic Compression: Constant Entropy, Compression of the gas due to the surroundings (environment) inputs work on the gas compressing it causing it to rise to hot temperature | 6 Ether resurrects to 9 Ether | Kala Rising |

The figure below synthesizes of all of the Thermodynamic Cycles described

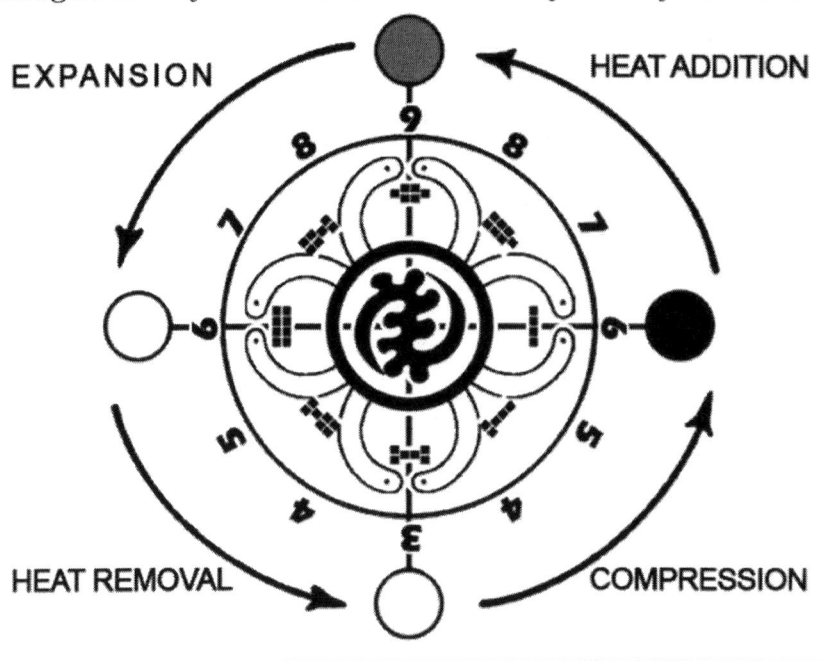

## 9 E.T.H.E.R.

Thermodynamics is one of the subjects that students at all levels from elementary to post-graduate find challenging. The allegories, parables, and analogies related to Thermodynamics found in African Theology and Theory provide effective mental tools for comprehending the subject of Thermodynamics. For **African Blacksmiths**, **Builders**, **Masons**, and **Craftsmen** it is necessary to apply the knowledge of Thermodynamics utilizing heat and temperature to expand, contract, and transform matter into new desired Creations. The Reality of "who and what your are", the purpose and meaning of life, "what happens after you die", and "death, resurrection, and reincarnation" can all be discovered by studying Thermodynamics. It is known that the essence of you is a substance called **DNA** or "**Deoxyribonucleic Acid**". Therefore, at your essence you are an **ACID**! Acids (you) chemically react with other substances and elements causing a transformation of the substances and elements from one thing to another. Fire and Electricity also cause chemical reactions. The Gas you breathe out is no different from the Exhaust or Smoke from a Fire. The excrement and waste you produce is no different from the ashes left over from a fire. Thus, what you call life is no more than the Thermodynamic cycles of the transfer of Energy in a series of Chemical Reactions. Your thoughts or Consciousness is the flow of electrons or electricity that occurs during the Energy transfer of the thermodynamic cycles during the chemical reaction of life. You are a Thermodynamic process, your purpose is to Create and Transform matter and energy via your Thermodynamic process, and you are in turn Transformed by Thermodynamic processes. Comprehend these concepts of Thermodynamics well! Each position on the thermodynamic diagram not only refers to a state of Energy, but also Mentality, Personality, and Sprit. Applying the science of Thermodynamics provides you with the ability to transform anything at will. The Practical Application of Thermodynamics Enables the Practitioner to figuratively create "Something from Nothing" which is indeed a Divine capability.

## 5.0. HYDRODYNAMICS

In Ancient Greek mythology, **Aether** was a deity who was the personification of the **upper Air**, **Atmosphere**, **Sky**, and **Heavens**. In some Ancient Greek mythological stories, Aether was the son of **Erebus** (**Darkness**) and **Nyx** (**Night**), while in other Greek mythological stories *Aether* was the son of **Chronos** (**Time**) and **Ananke** (**Destiny**). As a son of *Chronos*, *Aether* would have been a brother to **Poseidon**, the Greek deity of **Water**. In the Greek tradition, Poseidon was the father to another Greek deity of "**Wind**" named **Aeolus**. An interesting observation in Greek mythology is that the Greeks seemed to be confused about the progenitors and progeny of their own deities with various Greek stories giving different accounts for relationships between deities. Perhaps the confusion found in Greek mythology is due to the fact that the Greeks did not create their own system of mythology but rather adopted and transformed the mythology of Ancient Egypt as suggested in the book "**Stolen Legacy**" by **George G.M. James**. It is quite possible that the ambiguous relationship between the concepts of Air, Water, and Wind that were personified as *Aether, Posiedon,* and *Aeolus* in Greek mythology were derived from the more direct relationship between the concepts of **Gases** and **Moisture** that were personified as *Shu* and *Tefnut* in the Ancient Egyptian mythologies that predated Ancient Greece by thousands of years. At their essence, these various mythologies establish connections between Aether and **gases**, and **liquids**. In modern science, the movement and flow of Gases and Liquids is studied in the areas of **Aerodynamics**, **Hydrodynamics**, "**Fluid Dynamics**", and "**Fluid Mechanics**". In modern science, the term "Fluid" does not just refer to Liquids, but rather refers to states of matter including Gases,

# 9 E.T.H.E.R.

Liquids, Plasmas, and some solids that may flow like a fluid or take on the shape of a container like plastics, liquid metal, liquid glass, mud, or even **grains of sand** in an **Hour Glass**. The study of materials with both solid and fluid characteristics is called **"Rheology"** named from the Greek goddess and aunt of *Aether* named **Rhea** meaning **"flow"**.

**SHU**
Ancient Egyptian personification of "the Wind", Air, or Gases in Nature. He was the son of the Sun deity RE, and Husband of TEFNUT.

In Ancient Egypt, the personification of the wind, air, atmosphere, and gases in general was the deity named *Shu*. Shu was the husband of the goddess *Tefnut* who represented Moisture or Liquids in nature. *Tefnut* was depicted as a woman wearing the feline lioness mask. In Ancient Egyptian mythology, both *Shu* and *Tefnut* were children of the Sun deity **Atum-Re**. The relationship and personification of the wind and water can also be found in the West African **Orisha** tradition with the deity *Oya Iansan* representing "winds" and *Yemaya* representing "water".

# R.E. ENGINEERING

**TEFNUT**
Ancient Egyptian Principle representing the personification of "Moisture", Water, or Liquids in Nature. Daughter of the Sun deity RE, Wife of SHU.

The Ancient Egyptian story of **Shu** (Gases) and **Tefnut** (Liquids) being husband and wife and both being children of **RE** (Sun, Fire, or Heat) establishes a parable that shows relationships between modern scientific concepts like **steam**, **condensation**, and **evaporation** which are transitions between **gases** and **liquids** caused by **heat**. The application of the science of Thermodynamics as it applies to transitions between liquids and gases led to the development of one of the earliest forms of engines called the **Steam Engine**. Modern Science credits the invention of the steam engine to a Greek man named **Hero of Alexandria** who developed an object he called the **Aeolipile**. The Aeolipile, which literally means "the ball of the greek wind deity Aeolus", was a hollow metal ball with exhaust pipes coming out of opposite ends and being fed by a pool of water. When the metal ball of the Aeolipile was heated, the water would turn into steam and be forced out of the

## 9.E.T.H.E.R.

exhaust pipes causing the ball to spin. The Aeolipile demonstrated the fundamental scientific law that **"for every action, there must be an equal and opposite reaction"**. Hero of Alexandria was a member of an Ancient Greek organization called the **Atomist**. It has been suggested that the origin of the name of the Atomist group came from the Ancient Egyptian deity **Atum**. It could very well be that Hero of Alexandria was inspired by Ancient Egyptian science in the development of his Aeolipile. Even the name **"Hero"** is phonetically similar to **"Heru"** of ancient Egypt. The Ancient Egyptian wind and air deity *Shu* could have very well been the inspiration for the Greek deities *Aether* and *Aeolus*. One of the famous depictions of the Ancient Egyptian wind deity Shu is seated facing the **Ankh**. The Ankh is an Ancient African symbol called "The Key to Life", or "continuous life". The Ankh embodies the unification of dual principles and represents the flow of life giving fluid (called Semen) from the Male Principle at the bottom to the Female principle at the top which is **"the key to human life"** in the Ankh's Biological symbolism and the flow of Fluids (Gases and Liquids) due to temperature and pressure differences which is the **"key to fluid dynamics"** in the Ankh's Thermodynamic symbolism. Thus, rather than the "ball of *Aeolus* (Aeolipile)", the **Ankh of *Shu*** may have been the origin and inception of the concept of the steam engine.

**The Ball of *Aeolus* (Aeolipile)    The Ankh of *Shu* (Shu Ankh)**

www.AfricanCreationEnergy.com

# R.E. ENGINEERING

Hero of Alexandria also wrote about the Hydrodynamic principles of the **Siphon** in a book entitled **"Pneumatica"** which literally meant **"the application of breath, spirit, or soul"** coming from the Greek words **"*Pneuma*"** meaning "breath, spirit or soul" and the suffix "*-tica*" meaning "to apply". However, in the book entitled **"A History of Mechanical Inventions"** the author indicates that there is evidence that the Ancient Egyptians during the **18th Dynasty** used Siphons to draw oil out of **canopic jars** as early as 1500 BC. The Hydrodynamic knowledge of the Siphon may have been another Ancient Egyptian invention that influenced the Atomist of Greece and Hero of Alexandria.

Siphons are basically pipes or tubes that make use of Hydrodynamic principles to cause liquids to flow between two or more containers at different heights. The liquid flows through the Siphon because the liquids in the containers are at different heights and thus have a difference in gravitational pull, potential energy, and pressure.

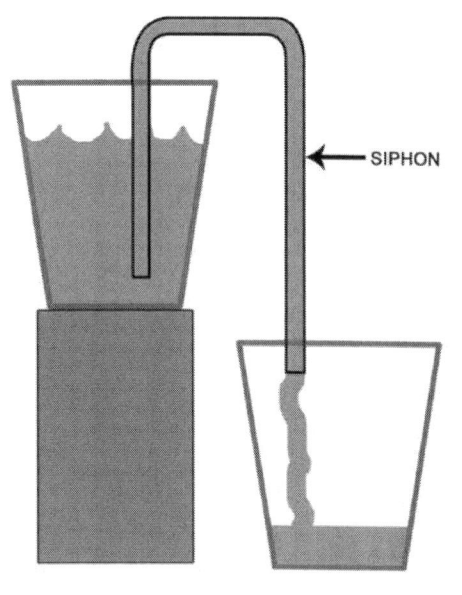

Just as the pressure in Hydrodynamic processes causes the movement of liquid in a siphon, similarly **Blood Pressure** causes the **circulation of blood** through your **veins** which enables **life**. As we mentioned earlier, both liquids and gases can flow as a "fluid". Much like liquids flow through a siphon due to a difference in pressure, gases flow causing wind due to a difference in pressure. In modern science, the equation that is used to mathematically model the relationship between

## 9 E.T.H.E.R.

**Pressure**, **Volume**, and **Temperature** of Air, or more specifically a Gas, is called the **Ideal Gas Law**. The Ideal Gas Law states that $PV=nRT$, where $P$ is "pressure", $T$ is "Temperature", $V$ is Volume, $n$ is the amount or quantity of Gas, and the constant $R$=8.314 J/K*mol. The Ideal Gas Law establishes a direct relationship between Temperature and Pressure. Thus, as Temperature increases, Pressure increases; and as Temperature decreases, pressure decreases. Wind is scientifically defined as the flow of gases on the macroscopic scale. Wind exists not only on the planet Earth, but also throughout outer space as **Solar Wind**. Wind occurs when there is a difference in pressure and/or a difference in temperature in a Gas causing the Gas to move or flow from the higher pressure area to the lower pressure area. On Earth, varying temperatures on the planet and varying pressures throughout the layers of the atmosphere create weather phenomenon.

Air, or more specifically, Gases have a somewhat "magical" quality in that if the gases are colorless and transparent, they cannot be seen with the naked eye, however the effects of the gas on physical objects can be observed. A strong enough wind can move or animate a stone making it seem "**alive**". It is no surprise that in modern science, the quality of a container being "**airtight**" is called a "**Hermetic Seal**" named for a famous "**magician**" named **Hermes Trismegistus** who was the synthesis of the Greek deity **Hermes** and the Egyptian deity **Tehuti** or **Thoth**. If you were shown an empty cup and asked "What is in the cup?" you would likely respond "**Nothing**". If you were shown a cup full of water and asked "Is there **something** in the cup" you would likely respond "Yes". However in the empty cup there is Air or Gases present. So in the mind of most people, Air or Gases is a form of "**Nothing**" and Water or Liquids is a form of "**Something**". In modern science, **Condensation** is the transformation of Gases into liquid. Thus, Condensation is a method of metaphorically transforming "**Nothing into Something**". The process of Condensation is what leads to **water vapor** in the form of gas condensing into droplets of water on **dust** particles in the creation and formation of **clouds** in the sky. It is interesting to note that the etymology of the word "Nine" comes from an Old English word "**nigonðe**" used to refer to a "**Cloud Nine**" also known as the **rain** producing **Nimbus cloud**. Just as clouds on Earth form as a result of condensation of water vapor and dust in the sky, **Interstellar Clouds** in outer space called **Nebula** are formed in a similar process of congealing **dust**, **hydrogen**, **helium**, and **ionized gases**.

Although the process of condensation can occur without human intervention in Nature, devices called **Condensers** can be used to transform gas into liquid by way of human intervention. **Thermo-electric** Condensers are found in both **Refrigerators** and **Air-Conditioners** used for cooling. Thermo-electric condensers are also used in devices called **Atmospheric Water Generators** which transform the water vapor in humid air into water available for human usage. Another type of non-electric Atmospheric Water Generator is a structure called an **Air Well**. Air Wells are massive Dome, Cone, or Pyramid shaped masonry structures which, due to their construction design, cause the moisture in humid air to condense into water within the structure. **Humidity** is the measure of water vapor in air, and the most humid places on Earth are usually close to the equator like many **African** countries. Thus, it is believed by some **Pyramidologist** that one of the purposes of the Pyramids constructed in Egypt was to serve as Air Wells to condense Air into Water (**marry Shu and Tefnut**) for the use of Egyptian people residing in the desert.

Conversely, just as condensation is the transformation of "Nothing into Something", the process of **Evaporation** is like the transformation of **"Something into Nothing"** where liquids actually evaporate or **"vanish into thin air"**. The Thermodynamic and Hydrodynamic processes involved in the Condensation and Evaporation of water is called the **Hydrologic Cycle**. Water can be transformed into a gas by the process of Evaporation, but water can also be transformed into gases through the process of **Electrolysis**. The chemical formula of water is **H₂O** meaning that one molecule of water is composed of 2 **Hydrogen** gas atoms and 1 **Oxygen** gas atom. Electrolysis of water occurs by using a DC (direct current) voltage source to separate water into Hydrogen gas and Oxygen gas. When the **Anode** (negative end) and **Cathode** (positive end) of a voltage source is placed in water, the Oxygen gas bubbles will be attracted to the Anode (negative end) and the Hydrogen gas bubbles will be attracted to the Cathode (positive end). The application of the **Electrolysis of water** can be used to **produce breathable Oxygen while underwater**, **create Hydrogen gas** for fuel or other flammable or explosive needs, and also to create **Alkaline drinking water**. Just like water can become ionized and Alkaline using electric voltage sources, **Air Ionizers** are electronic devices that ionize air making breathable air more conducive to the flow of Electricity and **Ether**.

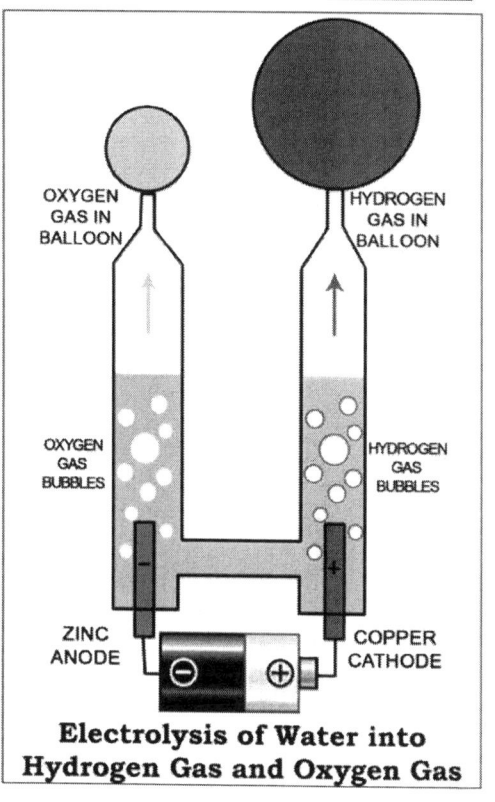

**Electrolysis of Water into Hydrogen Gas and Oxygen Gas**

www.AfricanCreationEnergy.com

## 6.0. ELECTROMAGNETIC RADIATION

In early modern Science, the term **"Luminiferous Ether"** referred to a **light-bearing ether** which was believed to be the medium needed for the **propagation of light**. The Scientist Albert Einstein also introduced the concept of **Relativistic Ether** in his theories about **light**. In addition to being the personification of the Air, the Greek deity **Aether** was also considered the personification of **"Light"**. Thus, there is a direct relationship between Ether and Light in Ancient Theology and Scientific Theory. In modern science, "**Light**" is a form of **Electromagnetic Radiation** that is visible to the human eye. Therefore, there is also a relationship between Ether and Electromagnetic Radiation in general.

Electromagnetic Radiation is a Fundamental Force in Nature. The range of frequencies of Electromagnetic Radiation is called the **Electromagnetic Spectrum**. "Light" is actually a very small portion of the entire Electromagnetic Spectrum. The various ranges of frequencies of the electromagnetic spectrum have been given different names based on their qualities. In order of lowest frequency to highest frequency, the names of the various ranges of frequencies of the electromagnetic spectrum are: **Radio Waves, Microwave Rays, Infrared Rays, Visible Light, Ultraviolet Rays, X-rays,** and **Gamma Rays**. In modern science, the smallest measurable unit (**quantum**) of "Light" or Electromagnetic Radiation is the **Photon**. Because "Light" or Electromagnetic Radiation can be explained as both a particle (photon) and a wave (spectrum or "Ray"), it is said that **"Light has a Dual Nature"** known as the **Wave-Particle Duality**.

# 9 E.T.H.E.R.

Light, or Electromagnetic Radiation, is caused by the movement of subatomic particles. In order to comprehend the origin and cause of Electromagnetic Radiation, we must first comprehend the structure of the **Atom**.

At the center of the Atom is a Nucleus composed of **Protons** and **Neutrons**. Protons and Neutrons are both composed of the fundamental particle know as the **Quark** with Protons being composed of two "Up Quarks" and one "Down Quark" and Neutrons being composed of one "Up Quark" and two "Down Quarks". The number of Protons an Atom has determines its "Atomic Number" or Element number on the Periodic Table.

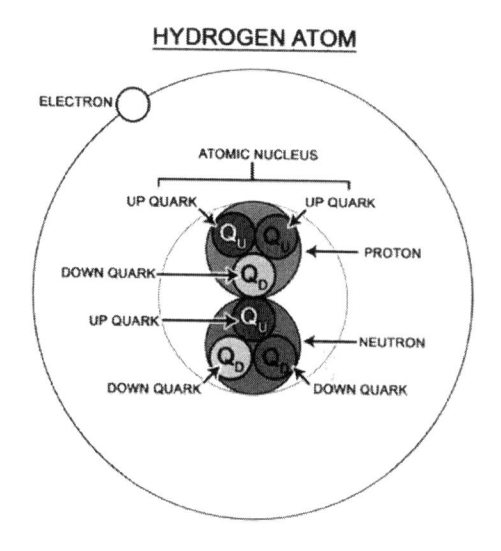

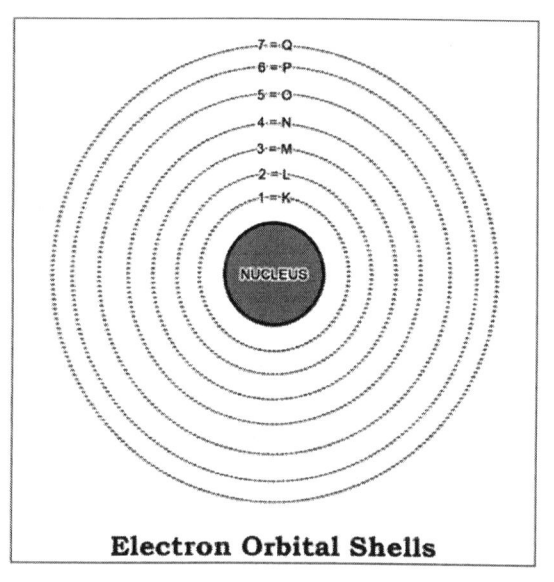

**HYDROGEN ATOM**

**Electron Orbital Shells**

Around the nucleus of the Atom are seven Orbital shells in which **Electrons** circle around the nucleus of the Atom like planets around the **Sun**. The electron Orbital shells are numbered from 1 to 7 or lettered from "K" to "Q". When an Atom absorbs Energy, the Electrons move from the lover Orbital shells near the Nucleus to the higher Orbital Shells away from the

www.AfricanCreationEnergy.com

## R.E. ENGINEERING

Nucleus.  When the Electrons go back to the lower Orbital shells near the Nucleus from the higher Orbital shells away from the Nucleus, a Photon of Electromagnetic Radiation is emitted from the Atom.  The movement of Electrons from each orbital shell in various Atoms produces the various frequencies of the Electromagnetic Radiation Spectrum.  The table below shows the predicted Electromagnetic Radiation emitted from an Atom of Hydrogen.

| Electron Orbital Shell FROM | TO | Frequency (THz) | Wavelength (nm) | Electromagnetic Radiation Type |
|---|---|---|---|---|
| 7 | 1 | 3297 | 91 | ULTRAVIOLET |
| 6 | 1 | 3191 | 94 | |
| 5 | 1 | 3158 | 95 | |
| 4 | 1 | 3093 | 97 | |
| 3 | 1 | 2913 | 103 | |
| 2 | 1 | 2459 | 122 | |
| 7 | 2 | 756 | 397 | UVA - Black Light |
| 6 | 2 | 732 | 410 | VISIBLE - Violet |
| 5 | 2 | 691 | 434 | VISIBLE - Blue-Indigo |
| 4 | 2 | 617 | 486 | VISIBLE - Cyan (Blue-Green) |
| 3 | 2 | 457 | 656 | VISIBLE - Red |
| 7 | 3 | 294 | 1020 | INFRARED |
| 6 | 3 | 274 | 1094 | |
| 5 | 3 | 234 | 1282 | |
| 4 | 3 | 160 | 1875 | |
| 7 | 4 | 138 | 2170 | |
| 6 | 4 | 114 | 2630 | |
| 5 | 4 | 74 | 4050 | |
| 7 | 5 | 64 | 4650 | |
| 6 | 5 | 40 | 7460 | |
| 7 | 6 | 24 | 12400 | |

X-Ray Electromagnetic Radiation with frequencies of at least 30000 THz is produced when electrons move between the lower Orbital Shells of elements heavier than Lithium.  When electrons move from the 3 shell (M shell) to the 1 shell (K shell)

## 9 E.T.H.E.R.

in heavy elements, a K-Alpha X-Ray is emitted. When electrons move from the 2 shell (L shell) to the 1 shell (K shell) in heavy elements, a K-Beta X-Ray is emitted. The emission of X-Ray Electromagnetic Radiation is called "**Characteristic X-Rays**". While most of the frequencies of Electromagnetic Radiation come from the movement of an atom's electrons between the various electron orbital shells, **Gamma Ray** Electromagnetic Radiation occurs due to changes in an atom's nucleus. Gamma Ray electromagnetic radiation is emitted during **Alpha decay**, **Beta decay**, or **Gamma decay** of a radioactive element. There are two types of Beta decay: Beta-plus decay and Beta-minus decay. Beta-plus decay occurs when an atomic Neutron becomes a Proton; that is when a Down Quark becomes and Up Quark. Beta-minus decay occurs when an atomic Proton becomes a Neutron; that is when an Up Quark becomes a Down Quark. **Neutrinos** are also emitted during Beta Decay.


The source and origin of Electromagnetic Radiation, or "Light" is the Atom. On Earth, the primary source of "Light" is the **Sun**. Thus, Sacred Geometry is observed in the fact that Macrocosmic **Solar Systems** resemble microcosmic **Atomic systems**. In fact, in the book "Stolen Legacy" by George G.M. James, it is suggested that the Greek word "**Atom**" was derived from the Ancient Egyptian Sun deity *Atum*. In Ancient Egyptian cosmology, Atum was one of several aspects of the "Sun" deity named *RE*. The Ancient Egyptian Sun deity *RE*, also spelled *RA*, was worshipped in the city called **Annu** which was known as **Heliopolis** to the Greeks, meaning "city of the Sun". The sun deity *RE* had a variety of aspects, each of which was represented by a different symbol. The aspects of *RE* named *Atum* and *Amun* were both personified as a man. *Atum* represented *RE* the sun in the **evening**, and *Amun* represented *RE* the sun at **night**.

# R.E. ENGINEERING

**RE**
Ancient Egyptian Sun God depicted as a Man (*Atum* or *Amun*) and a Falcon (*RE-Horakhty* or *Heru*)

The sun deity RE was merged with **Heru** or **Horus** to become **RE-Horakahty** symbolizing the two horizons of "morning and evening". RE-Horakahty was symbolized by the **falcon** or hawk. The aspect of RE the sun god which represented the morning sun was called **Khep-Re**, also spelled **Kephera** or **Khepri**. Khep-RE not only symbolized the morning sun but also symbolized **"Transformation"** and the movement of the sun across the sky. Khep-RE was symbolized by the scarab beetle. The word "*RE*" or "*RA*" itself meant

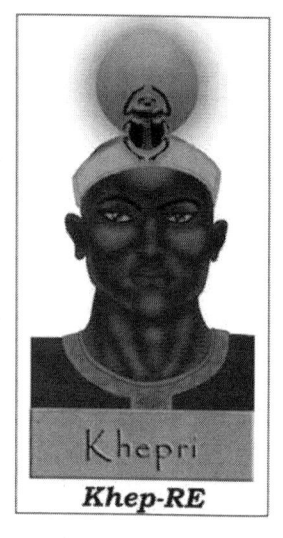

**Khep-RE**

"**creative power**" or "**creative energy**". In the story of "Aset and RE", or "Isis and RA", it is said that RE represents the sun at **High Noon**.

62 www.AfricanCreationEnergy.com

## 9 E.T.H.E.R.

Another aspect of the Sun deity RE was the sun disc called **Aton** or **Aten**. The Aton sun disc represented a synthesis of all the aspects of RE in one symbol. The Aton was depicted as a sun disc or a **winged-sun disc** with various arms or "Rays" emerging from the disc. It is interesting to note that the word "*RE*" or "*RA*" is phonetically similar to the word "Ray" and can also be found in the word "Radiation". Another interesting observation is that the sun disc Aton is occasionally depicted with 14 Rays emerging from it represented the "**14 Creative powers of RE**". Seven of the 14 Creative powers of RE were: **Heka** representing "magic", **Sia** representing intelligence, **Hu** representing utterance, **Maa** representing Sight, **Sedjem** representing hearing, **Ma'at** representing Order, and **Sa** representing a Universal cosmic force. In the book "The Sacred Magic of Ancient Egypt" by Rosemary Clark it is noted that the creative power of RE called **Sa** was equivalent to the concept known to the Greeks called **aether**. The sun was also personified in other African traditions as **Anyanwu** who represented to "eye of the Sun" in the **Igbo** culture and also as **Lisa**, the sun deity in **Dahomey** culture, who was the sibling to the moon deity **Mawu**, and offspring of the Orisha **Nana Buluku**.

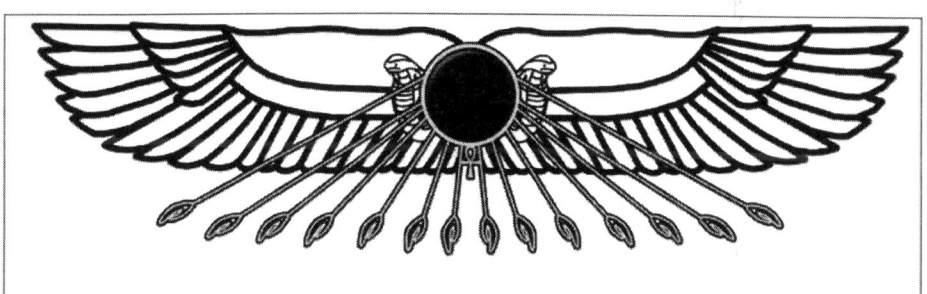

***Aton* or *Aten* the Sun Disc**
**With 14 "Rays" of Electromagnetic Radiation**
**as depicted by the Pharaoh Akhenaten**

In modern science we also find the prevalence of the number 7 and 14 when we study phenomena related to Electromagnetic

# R.E. ENGINEERING

Radiation. As mentioned earlier, there are 7 Electron Orbital shells 1) K, 2) L, 3) M, 4) N, 5) O, 6) P, and 7) Q. There are also seven frequency categories of the Electromagnetic Spectrum, and seven color frequency categories of the Visible Light portion of the electromagnetic spectrum. When we add the 7 Electromagnetic Radiation spectrum frequency categories with the 7 color frequency categories of Visible Light we get the number 14. Perhaps this is what was symbolically depicted by the Ancient Egyptians in the 14 rays emerging from the *Aton* sun disc and the 14 "creative powers" of *RE*.

The 7 Frequency Categories of the Electromagnetic Radiation Spectrum are:

| TYPE | Frequency | Wavelength |
| --- | --- | --- |
| Gamma ray | 100,000,000 THz + | 0nm TO 0.003 nm |
| X-Ray | 30,000 THz TO 30,000,000 THz | 0.01 nm TO 10 nm |
| Ultraviolet | 800 THz TO 30,000 THz | 10 nm TO 375 nm |
| Visible | 400 THz TO 800 THz | 380 nm TO 750 nm |
| Infrared | 0.3 THz TO 400 THz | 750 nm TO 1,000,000 nm |
| Microwave | 0.0003 THz TO 0.3 THz | 1,000,000 nm TO 1,000,000,000 nm |
| Radio | 0 THz TO 0.0003 THz | 1,000,000,000 nm + |

The 7 Color Frequency Categories of the Visible Light Portion of the Electromagnetic Radiation Spectrum are:

| Color | Frequency | Wavelength |
| --- | --- | --- |
| violet | 668–789 THz | 380–450 nm |
| indigo | 631–668 THz | 450–475 nm |
| blue | 606–630 THz | 476–495 nm |
| green | 526–606 THz | 495–570 nm |
| yellow | 508–526 THz | 570–590 nm |
| orange | 484–508 THz | 590–620 nm |
| red | 400–484 THz | 620–750 nm |

## 9 E.T.H.E.R.

The frequencies between 400 THz to 800 THz of the Electromagnetic radiation spectrum are the frequencies that Human beings can see with our eyes and hence this portion of the Electromagnetic radiation spectrum is called the "Visible Light Spectrum". Recall that the origin of Electromagnetic Radiation and each color frequency of Visible Light is due to the movement of electrons within an atom between different electron orbital shells. When all of the color frequencies of the Visible Light spectrum are combined, the product is called "**White Light**". White Light can be produced through **Thermal Radiation** in the visible light spectrum. Thermal Radiation is the electromagnetic radiation that is emitted by the motion of particles in matter. Not all Thermal Radiation occurs in the Visible Light range, but **the movement of sub-atomic particles in every object in nature indicates that it is indeed emitting Electromagnetic Radiation due to Thermal Radiation**. For example, Human Beings emit Electromagnetic Radiation in the Infrared range which can be seen using **Thermal imaging** technology. Certain objects can be "Heated" or Thermally Radiated to the point where they emit Visible Light or "White Light. The emission of Visible Electromagnetic Radiation due to Thermally Radiating or Heating an object which dramatically increases its temperature is called **Incandescence**. Examples of Visible Light or "White Light" due to Incandescent Thermal Radiation are the Sun, Incandescent Light Bulbs, and Fire. In modern science, the name for an ideal absorber and perfect emitter of "White Light" due to Thermal Radiation is called a **Black-Body** and the Incandescent electromagnetic radiation emitted by a Black-Body is called **Black-Body Radiation**. However, Visible Electromagnetic Radiation or "White Light" can also be emitted in processes that do not involve heating an object to dramatically increase its temperature called **Luminescence**.

# R.E. ENGINEERING

Because Luminescence does not dramatically increase the temperature of the light emitting object, it is considered a form of **"Cold-Body Radiation"**. Although Luminescence does not dramatically increase the temperature of the light emitting object, Thermal Radiation is still occurring in the process. There are a multitude of processes that can emit visible light by way of luminescence. Some of these processes include:

1. **Photo Luminescence** – visible light emitted due to the absorption of photons. This is the process used in **Fluorescent Lights**

2. **Electrical Luminescence** – visible light emitted due to an electric current passing through a substance (**LED**)

3. **Mechanical Luminescence** – visible light emitted due to a mechanical force acting on an object

4. **Chemical Luminescence** – visible light emitted due to a chemical reaction

5. **Biological Luminescence** – visible light emitted by a living organism

6. **Crystal Luminescence** – visible light emitted during the process of crystallization

7. **Sonar Luminescence** – visible light emitted when sound waves cause bubbles to collapse within a liquid

**Luminescent** and **Incandescent** light sources can be compared to the **"Great Light"** and the **"Lesser Light"** spoken of in the Christian Bible in Genesis chapter 1 verse 16, or to the **Noor** and **Naar** spoken of the Islamic Quran as the substance which **Angels** and **Jinns** are made of respectively.

White Light is a **"chaotic light"** in that it is formed by the combination of different frequencies which are caused by

## 9 E.T.H.E.R.

different electrons moving between different electron orbital shells seemingly "**doing their own thing**". Conversely, the electromagnetic radiation or Light emitted from a **LASER** is an "**orderly**" or "**coherent light**". The Electromagnetic Radiation emitted from a **LASER** is all of one frequency, one color, and thus comes from electrons moving between the same orbital shells all "**doing the same thing**". The word "LASER" is an acronym standing for "Light Amplification by Stimulated Emission of Radiation". Currently, modern scientist can construct a LASER with various materials each of which emit electromagnetic radiation from as low as Radio wave frequencies to as high as X-Ray frequencies. The table below shows different materials that can be used to make a LASER.

| Laser Type | Frequency (THz) | Wavelength (nm) | Electromagnetic Radiation Emitted |
|---|---|---|---|
| Argon fluoride | 1554 | 193 | Ultraviolet |
| Krypton fluoride | 1210 | 248 | Ultraviolet |
| Xenon chloride | 974 | 308 | Ultraviolet |
| Nitrogen | 890 | 337 | Ultraviolet |
| Helium cadmium | 680 | 441 | Visible - Violet |
| Argon | 615 | 488 | Visible - Blue |
| Copper Vapor | 588 | 510 | Visible - Green |
| Helium neon | 552 | 543 | Visible - Green |
| Krypton | 528 | 568 | Visible - Yellow |
| Gold Vapor | 478 | 627 | Visible - Red |
| Ruby (CrAlO$_3$) | 432 | 694 | Visible - Red |
| Erbium | 199 | 1504 | Near Infrared (NIR) |
| Hydrogen Fluoride | 111 | 2700 | Near Infrared (NIR) |
| Carbon dioxide - $CO_2$ | 28 | 10600 | Far Infrared |

# R.E. ENGINEERING

A LASER that emits electromagnetic radiation in the Infrared and Microwave range like the **$CO_2$ LASER** is able to **cut through steel** because Infrared radiation is Heat which has the ability to melt through its target. The first LASER ever constructed primarily emitted electromagnetic radiation and heat in the Radio and Microwave range. The first LASER was actually called a **MASER** standing for "Microwave Amplification by Stimulated Emission of Radiation". The MASER was constructed in the 1950s and the first LASER which emitted electromagnetic radiation in the visible light spectrum was constructed in the 1960s. Recently, in the year 2009, the science of amplifying various phenomena by stimulated emission of radiation has been applied to create the **SPASER** (standing for Surface Plasmon Amplification by Stimulated Emission of Radiation) and the **SASER** (standing for Sound Amplification by Stimulated Emission of Radiation).

Because the various color frequencies of visible light combine to produce "White Light", a **"White Light LASER"** requires the combination of multiple LASER sources of each color frequency. Whereas the "White Light" that is emitted from an Incandescent source like

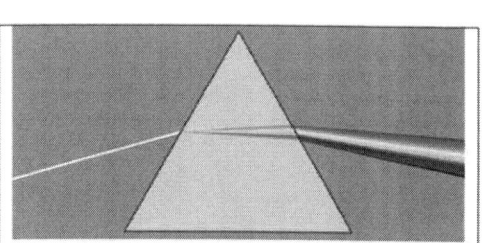

**Prisms can separate White Light into its component colors, or can combine color frequencies from LASERs to produce White Light through refraction**

the Sun can bend and be **refracted** and separated into each color frequency using a **Prism**, the different color frequencies of several LASERs can bend and be refracted and combine to form a "White Light LASER" using a **Prism**. Thus, the Prism is a tool that one can use to literally **"Bend Light"** or be an **"Ether Bender"**.

## 9 E.T.H.E.R.

Another "Ether Bending" tool that can be used to bend or **refract** light or electromagnetic radiation is the **Lens**. When electromagnetic radiation is refracted it changes **speed** and **direction**. When the lens has a **concave** shape, it causes the light passing through it to **diverge** and disperse. When the lens has a **convex** shape, it causes the light passing through it to **converge** and be **concentrated** at a **focal point**. Convex lens are used for **magnifying glasses** and can also be used to make a **"Burning Glass"** which concentrates the rays of electromagnetic radiation from the sun onto a target to be heated or **burned**. The oldest "Burning Glass" found to date is a 3000 year old lens made of **Quartz crystal** that was found in the city **Nimrud** of Ancient **Assyria**. The science of using convex lenses to concentrate light can also be done with mirrors. Convex mirrors which reflect and refract light were used by the Greek Mathematician **Archimedes** who developed a **"Heat Ray"** and also by the Ancient Egyptians to illuminate the inside of Pyramids and tombs.

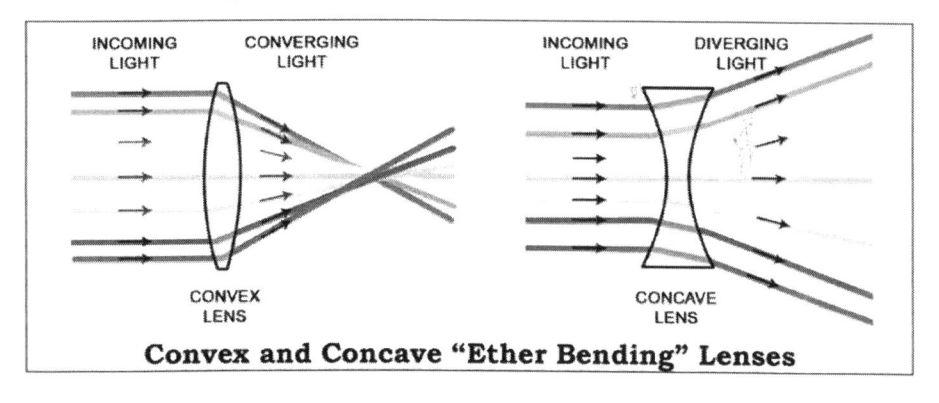

**Convex and Concave "Ether Bending" Lenses**

The way that white light is separated into the colors of the visible light spectrum through a prism is similar to the way the water vapor and clouds in the Earth's atmosphere separates the white light from sun light into the **Natural Holograms** called **Rainbows**. The sun generates all frequencies of the electromagnetic radiation spectrum. However, the Gamma Ray frequencies generated by the sun remain within the sun and are not emitted into space. The speed of electromagnetic radiation in a vacuum is 299,792,458 meters per second or 186,282 miles per second. As electromagnetic radiation or light travels through mediums or materials more dense than a vacuum, then the speed and direction of the electromagnetic radiation or light can change. The measure of how electromagnetic radiation travels through a **medium** is called the **Refractive Index**. For a **Vacuum**, the Refractive index is 1, for **Air** the refractive index is 1.000293 and for **Water** the refractive index is 1.333. The speed of light will slow down as light travels through mediums more dense that a Vacuum or with a refractive index greater than 1. However, for **Plasma**, the refractive index is less than 1. Recall that the Sun is plasma. An interesting phenomenon to note is that since the Sun is plasma and has a refractive index less than 1, then the X-Ray frequencies travelling within the Sun are able to travel **"faster than the speed of light"** (exceed the phase velocity of electromagnetic radiation in a vacuum). When the X-Ray frequencies are emitted from the Sun into the vacuum of space, they slow down to "the speed of light". Ultraviolet Rays, Visible Light, and Infrared Rays from the Sun make contact with the Earth's atmosphere. UVC Ultraviolet electromagnetic radiation is completely absorbed by the Earth's Atmosphere and does not make contact with the Earth's surface. UVB Ultraviolet electromagnetic radiation is partially absorbed by the Earth's Atmosphere and only partly reaches to Earth's surface. People living on parts of the Earth which receive UVB rays from the sun have developed darker complexions to absorb these ultraviolet electromagnetic rays within the skin. Considering that electromagnetic radiation from the sun does not originate on Earth, but has genetically influenced some people living on Earth, we can consider the **"People of the Sun"** whose

## 9 E.T.H.E.R.

DNA and melanin has been affected by the Sun empirical examples of **Extra-Terrestrial beings**.

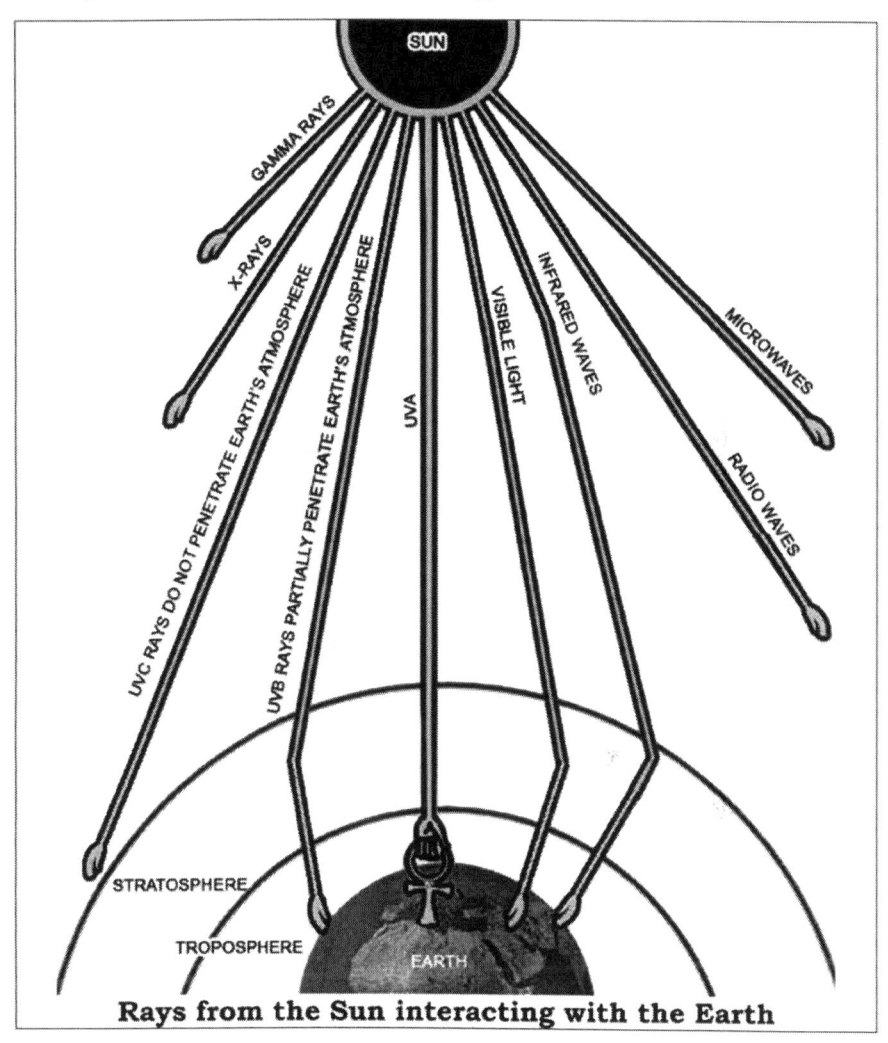

**Rays from the Sun interacting with the Earth**

Electromagnetic radiation from the sun also affects **hair texture**. The texture of a person's hair is determined by the shape of the **hair follicle**. A round hair follicle produces straight hair and a more oval or slit shaped hair follicle produces coiled hair. The shape of the hair follicle is determined by temperature of the person's genetic place of origin on the planet and the exposure of electromagnetic radiation in the region.

Hair color and hair texture is not unique to a person's Race. Although most people of the same race have similar complexions

| HAIR TYPE | STRAIGHT | WAVY | CURLY | COILED |
|---|---|---|---|---|
|  | 6 ETHER | 7 ETHER | 8 ETHER | 9 ETHER |
| HAIR FOLLICLE | ○ | ○ | ⊙ | ⊖ |

and hair textures, there are people with very dark complexions with naturally straight hair and there are people with very light complexions with naturally coiled hair. There are also people with very dark complexions that have naturally light hair and people with very light complexions that have naturally dark hair. It is the genetic exposure to electromagnetic radiation that determines a person's complexion and hair texture.

Just as the sun emits electromagnetic radiation in the form of Sunlight to the Earth, the Stars throughout the Universe also emit electromagnetic radiation in the form of Starlight. Just as **Sunlight** is absorbed by the plants and people of Earth, **Starlight** is also absorbed by the plants and people of Earth. Therefore, just as there are "People of the Sun" there are also **"People of the Stars"**. In fact our Sun is a Star. Electromagnetic radiation in the form of Sunlight and Starlight directly determine characteristics of our genetics through absorption by our bodies and indirectly through absorption by the fruits, vegetables, and food we eat. When we observe certain colors in Nature, it is indicative of certain elements. We must distinguish between the "color" of a Light Source and the "color" of an object that does not emit light. The color that is seen in these two cases occurs in two different ways. The color of a light source occurs from the movement of electrons between orbital shells creating a "color of light" from a frequency or a combination of frequencies. The color of an object occurs when light from a source strikes an object, and the object absorbs some frequencies of the light, and reflects other frequencies of the light. The frequencies that are not absorbed by the object determine the "color" of the object you see. For example, if you see a "green plant", that means the plant absorbs all of the color frequencies of the visible electromagnetic radiation spectrum except for the green frequency. Conversely, if you see a "Green LASER" that means the LASER emits only the green frequency of the visible electromagnetic radiation spectrum. If you see a "white wall", that means the wall does not absorb any of the color frequencies of the visible electromagnetic radiation spectrum. Conversely, if you see a "White Light" that means the light emits all of the color frequencies of the visible electromagnetic radiation spectrum. If you see the black fur of a "black cat", that means the black fur of the black cat absorbs all of the color frequencies of the visible electromagnetic radiation spectrum. Conversely, if you see a "**Black Light**", that means the source does not emit any color frequencies in the visible electromagnetic radiation spectrum. Therefore we comprehend that just as **Light** is a form of electromagnetic radiation, **Darkness** is also a form of electromagnetic radiation. **Brightness** is when we see a source or object that appears to be radiating or reflecting visible electromagnetic radiation, and **Darkness** is when we see a source or object that does not appear to be radiating or reflecting visible electromagnetic radiation. However, although things that we consider "Dark" do not emit or reflect visible electromagnetic radiation, there still may be electromagnetic radiation present in the form of Microwaves, Radio waves, Infrared ray, Ultraviolet Rays, X-rays, or Gamma Rays. Dark is the absence of visible electromagnetic radiation frequencies, but **Dark is NOT the absence of ALL electromagnetic radiation frequencies**. Even the slowest frequencies of electromagnetic radiation that we see as "Dark" still contain some electromagnetic radiation in the form of **Ultra-low frequency Waves** (emitted by Geomagnetic pulsations) between **0.001 Hz** and 3 Hz. An interesting paradox is that the word "Bright" and the word "Black" both have the same etymological origin. On a **"Pitch Black"** Night or in a **"Pitch Black"** room which is completely "Dark" to your eyes, there is still electromagnetic radiation present in frequencies outside of the visible light spectrum. This can be verified by using **Infrared Goggles** in "Pitch Black" dark environments to detect the Infrared electromagnetic radiation present. Most of the electromagnetic radiation frequencies that we are familiar with and consider "Dark" are Infrared, Microwave, and Radio waves. When Visible light frequencies shine into environments containing only Infrared, Microwave, and Radio frequencies, we say that the **"light has been turned on"**. When the visible light source is blocked or when it stops shining into environments containing only Infrared, Microwave, and Radio frequencies, we call this **"shade"**, **"shadow"** or say that the **"light has been turned off"**. However, if we were to shine Infrared, Microwave, Radio wave, or X-Ray, or Gamma ray frequencies into an environment containing visible light frequencies, we would not see "darkness". But, there is a portion of the Ultraviolet electromagnetic radiation spectrum that could be emitted into an environment containing visible light frequencies and we would see it as **"turning on the dark"**. Part of the UVA portion of the Ultraviolet electromagnetic radiation spectrum is visible to us and part of it is not visible to us or appears dark. If a UV LED Flashlight that emits electromagnetic radiation in the UVA range of the Ultraviolet electromagnetic radiation spectrum is shined onto the wall of a room lit with visible light, we will be able to observe a **"Flash Dark"** or **"End<u>ark</u>enment"** (opposite of en<u>ligh</u>tenment). Because

## 9 E.T.H.E.R.

"light" can be emitted into "dark" environments and still be perceived, but most frequencies that we see as dark cannot be emitted into environments full of "visible light' and still be perceived, we actually find that Christian Bible verse of *"the light shining in the dark and the dark comprehended it not"* (John 1:5) is incorrect, and the reality is that *"Dark can shineth into the Light, and the Light comprehended it not"*. Thus, what we perceive as "dark" and "light" on the electromagnetic radiation spectrum is divided with "light" being the visible light spectrum, and "dark" being Infrared and slower frequencies or Ultraviolet and faster frequencies. Therefore there are two "ranges" for the electromagnetic radiation that we perceive as "dark". These two types of darkness were personified in Ancient Egypt as the two members of the **Ogdoad** or **Khemenu** named **Kek** and **Keket**. Methods of comprehending the difference between electromagnetic radiation seen as "dark" to the human eye versus electromagnetic radiation that we see as transparent is studied in the fields of **Photometry** and **Radiometry**. The field of Radiometry is the study and measure of electromagnetic radiation in general. Photometry is the study and measure of electromagnetic radiation as perceived by the human eye. Every quantity and metric in the field of Radiometry has an analogous quantity and metric in the field of Photometry. The difference between **"Photometry versus Radiometry"** is like the difference between **"Perception versus Reality"** respectively. In the field of Photometry, the electromagnetic radiation frequencies that we see as "dark" can be determined from the **"Luminosity Function"** which mathematically shows the sensitivity of the human eye to various frequencies of electromagnetic radiation. As shown by the **Luminous Flux** of the Luminosity Function, humans perceive "Infrared and slower frequencies" just as "Dark" as "Ultraviolet and faster frequencies". However, in Radiometry we comprehend that it is possible for the frequencies that we perceive as dark in Photometry to have just as much if not more

www.AfricanCreationEnergy.com

# R.E. ENGINEERING

**Radiant Energy** than frequencies that we see as "light" in Photometry based on **Radiant flux**.

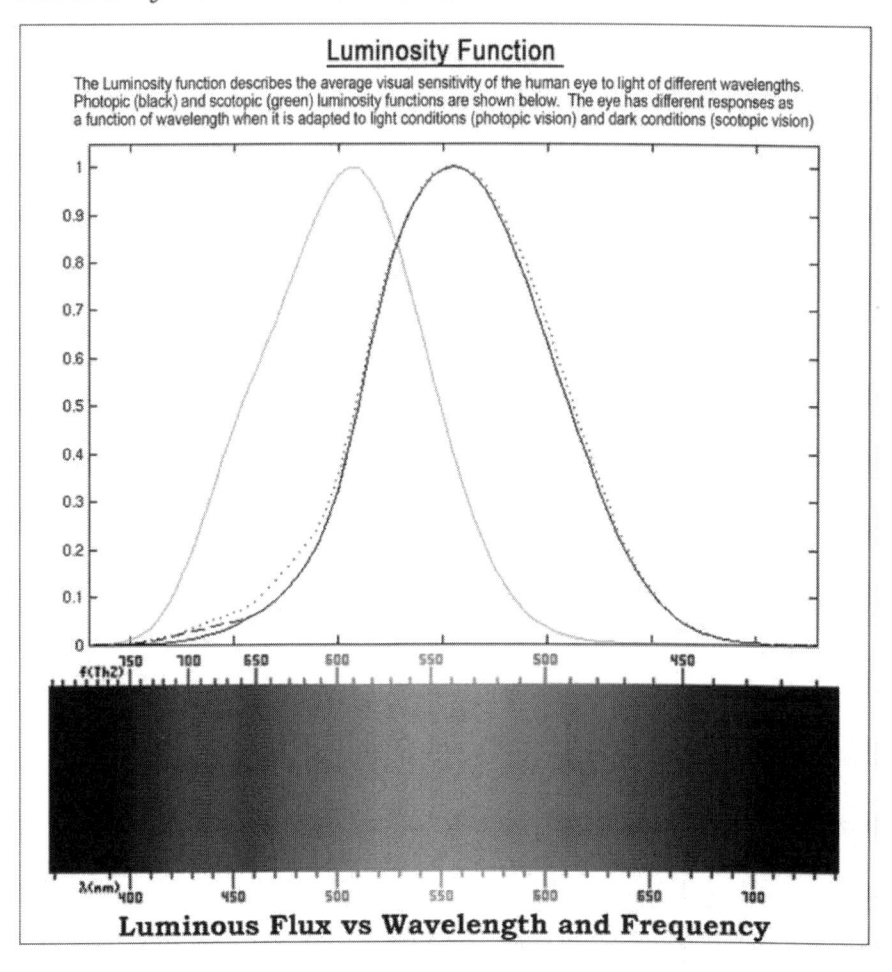

**Luminosity Function**

The Luminosity function describes the average visual sensitivity of the human eye to light of different wavelengths. Photopic (black) and scotopic (green) luminosity functions are shown below. The eye has different responses as a function of wavelength when it is adapted to light conditions (photopic vision) and dark conditions (scotopic vision)

**Luminous Flux vs Wavelength and Frequency**

Therefore, "Light" and "Dark" are both relative terms for electromagnetic radiation. For example, **Sunspots** are spots that occur on the surface of the sun and appear "Dark" relative to the rest of the "light" sun. However, if the "Dark" sunspot were removed from the sun and compared to other "Light" sources, the Sunspot would appear as "bright" as lightning. The quantity of these "dark"

## 9 E.T.H.E.R.

Sunspots on the surface of the sun is what determines the 11 year **Solar Cycle** or "**Sun Cycle**" or "**RE Cycle**".

Just as the Sun emits electromagnetic radiation, the thought processes of your brain also emit electromagnetic radiation. However the electromagnetic radiation emitted by your brain's thought processes occurs in the slowest "Radio wave" frequency of the electromagnetic radiation spectrum. These Extremely Low Frequency electromagnetic radiation waves emitted from your brain are abbreviated as **ELF** waves. The table below shows that different brain frequencies correspond to different mental states and thought patterns.

| TYPE | FREQUENCY (Hz) | MENTAL STATE |
|---|---|---|
| DELTA | 0 Hz - 4 Hz | Sleep or death |
| THETA | 4 Hz - 8 Hz | Drowsiness, idling, or thinking |
| ALPHA | 8 Hz - 13 Hz | relaxed, reflecting, meditation, inhibition |
| BETA | 13 Hz - 30 Hz | alert, working, active, busy or anxious thinking, active concentration |
| GAMMA | 30 Hz - 100 Hz+ | Cross-modal sensory processing (perception that combines two different senses, such as sound and sight), short term memory matching of recognized objects, sounds, or tactile sensations |
| MU | 8 Hz - 13 Hz | imitating, mirroring, or copying another person's behaviors |

It has been shown experimentally that a person's brainwaves can be affected, influenced, and changed to correspond to a different frequency if the person is exposed for a long enough period of time to the different frequency. In electrical systems, the transfer of a frequency from one object to another is a form of **Resonant Energy** transfer.

## 7.0. RESONANT ENERGY

In modern science, the term "**Resonance**" is used to refer to things that vibrate or oscillate at a certain **frequency**. Resonance occurs in anything that vibrates. Examples of resonance can be found in Electrical systems, Mechanical systems, Optical systems, Acoustic systems, Atomic systems, Solar systems, and Electromagnetic Radiation. The etymology of the word "Resonance" comes from the prefix "**Re-**" meaning "again" and the suffix "*-sonance*" meaning "sound". Thus, the etymological sense of the word "Resonance" is "to Resound", "to sound again", or the "**sound of Re**".

In the Heliopolis cosmology of Ancient Egypt, the "sound" or "utterance" emitted from the Sun deity **RE** was called **Hu** (*Hu-Re* or *Heru* or *Haru* "Horus"). The deity *Hu* was the personification of Creative Utterance, tone, pulsation, sound, and Vibration. *Hu* is comparable to the concept of the creative tone in Hiduism called **Aum** or **Om**. In the Memphite cosmology, it is said that the African Creation deity **PTAH** uttered the creative tone *Hu* to initiate creation. The Ancient Egyptians recognized the great Sphinx of Giza as a physical representation of *Hu*. The Ancient Egyptian name for the great Sphinx of Giza

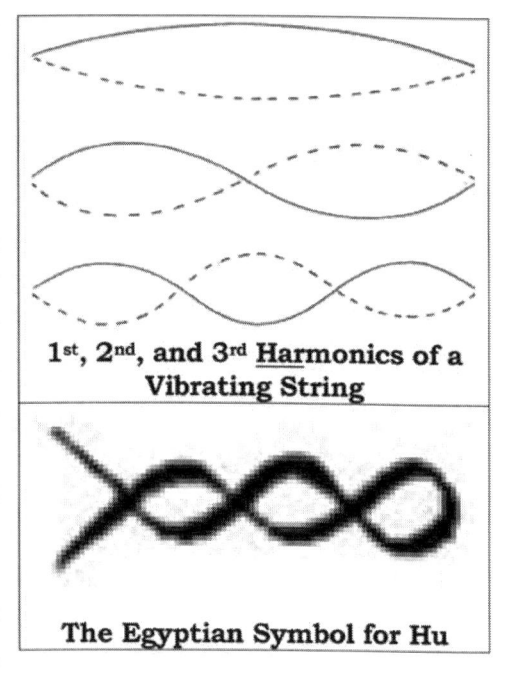

1st, 2nd, and 3rd **Harmonics of a Vibrating String**

**The Egyptian Symbol for Hu**

## 9 E.T.H.E.R.

was **_Har-em-akhet_** meaning "Horus in the Horizon". The word "**_Har-em-akhet_**" was mispronounced by the Greeks as "**Harmachis**" which is phonetically similar to the word "**Harmonics**" which is used in modern science to refer to a type of vibration or oscillation that can cause Resonance. Sound can also be used to **levitate** matter into the air through a process called **Acoustic Levitation**. Much like the **Photon** is considered the particle for Electromagnetic Radiation or "**Light**", the **Phonon** is considered the particle for **Sound**.

In electrical systems or "**Ether systems**" the transfer of Resonant Energy can be achieved through a process called **Resonant Inductive Coupling**. The book "**P.T.A.H. Technology: Engineering Applications of African Sciences**" by **African Creation Energy** describes the function of the electrical components called the **Resistor**, **Capacitor**, and **Inductor** in greater detail. The Inductor is a coil of wire (which resembles "**coiled 9-Ether Hair**" as shown in the previous chapter) which can be constructed in the shape of an **Ankh**. Inductors can also be used to heat an electrically conductive object like metal through a process called **Electromagnetic Induction Heating**. When metal is placed in the core of an Inductor and a high frequency Alternating Current (AC) is passed through the wire of the inductor, the metal will begin to heat up and thermally radiate. Electromagnetic Induction Heating can be used by **Blacksmiths** to weld, shape, and forge metal and it can also be used by **Masons** to shape, and manipulate stone.

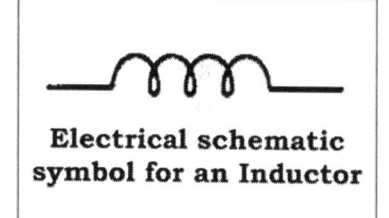

Electrical schematic symbol for an Inductor

Resonant Inductive Coupling is a form of **Wireless Energy Transfer** which transmits "Ether" or **electrical energy** between two Inductor **Solenoid** Coils which are tuned to resonate at the same frequency. If one coil is connected to an AC voltage source and a tuning capacitor, and a second coil is connected to a capacitor and an output load, then the electrical energy in the magnetic field generated by the first coil can be picked up by the second coil if the two coils are in close enough proximity and tuned to the same resonant frequency. Instructions on how to build an **"Ankh Wireless Energy Transfer Resonant Transformer"** are provided as an experiment in the next section of this book. The Wireless Energy Transfer Resonant Transformer is a type of **LC Circuit** which is one of several types of **Tuned Circuits** used in electronics. Tuned Circuits utilize various configurations of Resistors (symbolized by **"R"**), Capacitors (symbolized by **"C"**), and Inductors (symbolized by **"L"**) to build electrical circuits at various **Resonance Frequencies**.

Another type of Tuned electrical circuit is the **RC Circuit**. Whereas the LC Circuit can enable the wireless transfer of electrical energy through the air, **Radio Control** or **"RC"** can enable the wireless transfer of **Radio frequencies** of electromagnetic radiation through the air. Just as different radio wave frequencies in the brain correspond to different mental states or activities, the various frequencies wirelessly transferred through the air to Radio Controlled devices like toy cars, trucks, and helicopters correspond to different instructions or actions. When the wireless signal has a frequency in the Infrared range of the electromagnetic spectrum, then the term **"Remote Control"** is used. Thus, Resonant Energy or **Resonance** is a method for "consciousness" or **Reason** to be transmitted through **Ether**.

# 9 E.T.H.E.R.

## 8.0. ETHER EXPERIMENTS TO EXPERIENCE EVIDENCE

**EXPERIMENT 1: Calculate the Antimatter emitted from an African Yam**

Potassium-40 is a naturally occurring isotope that is unstable and radioactively decays. The radioactive half-life of Potassium-40 is $3.938 \times 10^{16}$ seconds. When Potassium-40 radioactively decays, 1 in every 100,000 decays emits an **Anti-Electron** (Positron) of **Antimatter**. For every 1,000,000 atoms of Natural Potassium, about 117 atoms are Potassium-40. Therefore, substances dense with Potassium will likely contain Potassium-40 and will also occasionally emit Antimatter in the form of an Anti-Electron. The atomic mass of Potassium is 39.1 g/mol.

1 Banana contains 0.450 grams of Natural Potassium

= $0.450 \text{ g} \times (6.02 \times 10^{23}) \div 39.1 = 6.93 \times 10^{21}$ atoms of Potassium

= $(117 \div 1,000,000) \times 6.93 \times 10^{21} = 8.11 \times 10^{17}$ atoms of Potassium-40

= $8.11 \times 10^{17} \div 3.938 \times 10^{16} = 20.58$ decays of Potassium-40 per second

= $(100,000 \text{ decays} \div 20.58 \text{ decays per sec.}) \div 60$ sec. per min.

= **Every 81 minutes an Anti-Electron Antimatter particle is emitted from a Banana**

Use the formulas provide above to calculate how long it will take for the African Yam, Sweet Potato, and Plantain foods rich in Potassium to emit an Anti-Electron particle of Antimatter.

| Food | grams of Potassium | Atoms of Potassium | Atoms of Potassium-40 | Decays of Potassium-40 per Second | Minutes until Anti-Electron is emitted |
|---|---|---|---|---|---|
| Banana | 0.450 | $6.93 \times 10^{21}$ | $8.11 \times 10^{17}$ | 20.58 | 81 |
| African Yam | 0.694 | | | | |
| Sweet Potato | 0.610 | | | | |
| Plantain | 0.534 | | | | |

# R.E. ENGINEERING

### EXPERIMENT 2: Create a Staff of PTAH to convert DC to AC

**WARNING:** This experiment can be dangerous. Proceed with caution! Minors should get the permission and supervision of a parent or guardian prior to performing this experiment. The Author is not liable for accidents that occur while performing these experiments.

**Materials Needed:**

- 60 meters (200ft) of 24 to 30 gauge magnet wire
- Toroid iron ring
- tuning fork
- 9 Volt Battery
- Alligator clips
- Neon Light (requiring 120 V 60 Hz AC)

**Instructions:**

1. Use the picture diagram to assist you with making the connections. Wrap the wire around the Toroid iron core 10 times on one side making sure to not overlap any of the turns of wire. Leave two ends of the wire free.
2. Wrap another wire 150 times around the other side of the Toroid iron core making sure not to overlap any of the turns of wire. Connect the two ends of the wire wrapped 150 times to the ends of the AC Neon Light.
3. Connect one end of the wire wrapped 10 times to one terminal of the 9 volt battery. Place the other end of the wire very close to the Tuning fork so that the end of the wire is touching the tuning fork.
4. Connect the other terminal of the 9 volt battery directly to the Tuning fork.
5. Tap the tuning fork so that it starts to vibrate. As it vibrates it will make contact with the wire sending pulsed DC voltage to the transformer and lighting the light.

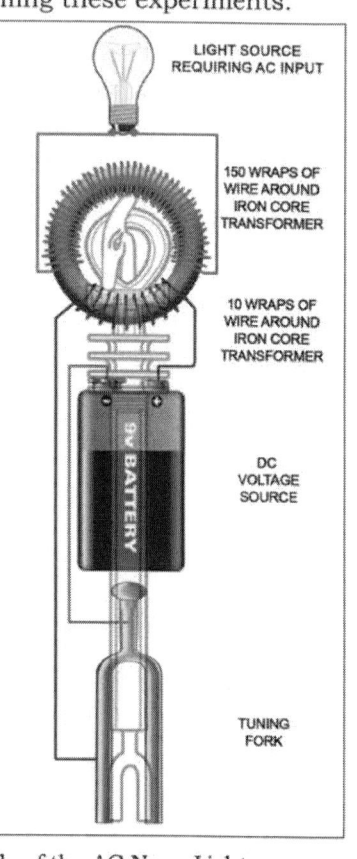

# 9 E.T.H.E.R.

### EXPERIMENT 3: Create a Solar Barque of RE

**WARNING:** This experiment can be dangerous. Proceed with caution! Minors should get the permission and supervision of a parent or guardian prior to performing this experiment. The Author is not liable for accidents that occur while performing these experiments.

**Materials Needed:**
- Sardine Can or plastic bottle
- tee-warmer candle
- 3 mm brass pipe
- Pliers
- Screw driver

**Instructions:**
1. Use the pliers to bend the brass pipe around the handle of a screw driver so that there are at least two to three turns in the coil and so that the ends are the same length
2. Remove the lid from the sardine can and punch two holes in the sardine can. If you are using a plastic bottle, cut the bottle in half, then punch holes in the bottle as previously described and depicted below.
3. Place the ends of the coiled brass pipe through the holes.
4. Put a tee-warmer candle under the coiled part of the brass pipe
5. Light the candle and the thermodynamic processes will cause the little Steam-boat to move

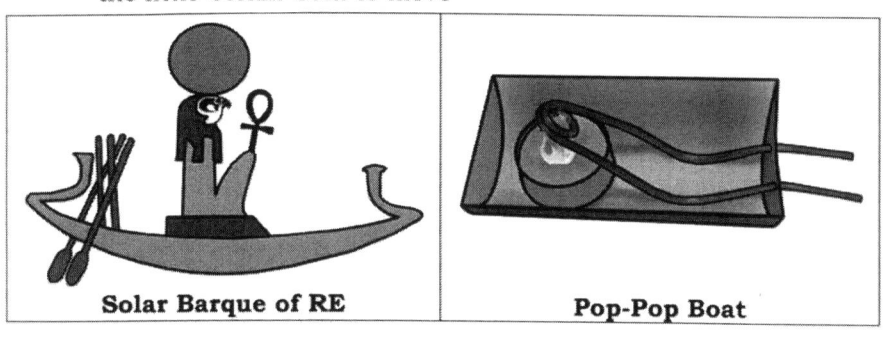

| Solar Barque of RE | Pop-Pop Boat |

# R.E. ENGINEERING

## EXPERIMENT 4: Create a Hot-Air Balloon

**WARNING:** This experiment can be dangerous. Proceed with caution! Minors should get the permission and supervision of a parent or guardian prior to performing this experiment. The Author is not liable for accidents that occur while performing these experiments.

**Materials Needed:**
- Scissors
- Scotch Tape
- Plastic Produce Bag
- 4 Birthday Candles
- 6 Straws
- Aluminum Foil
- Fire source

**Instructions:**
1. Cut the seam off of the top of the plastic bag
2. Lay the plastic bag out flat and tape the top of the bag closed where the seam was cut off
3. Tape the straws together forming an X
4. Tape each end of the straw X to the plastic bag
5. Tape a small square piece of aluminum foil to center of the straw X
6. Melt the bottom of the 4 candles to the aluminum foil
7. Light the candles
8. With the candles lit, hold the plastic bag up and open with your hands. The hot air will fill the plastic bag making it lighter than the surrounding air and the Hot-Air balloon will rise based on the principles of Thermodynamics.

## 9 E.T.H.E.R.

### EXPERIMENT 5: Create a Shu Ankh (Aeolia Pile)

**WARNING:** This experiment can be dangerous. Proceed with caution! Minors should get the permission and supervision of a parent or guardian prior to performing this experiment. The Author is not liable for accidents that occur while performing these experiments.

**Materials Needed:**
- Soda Can
- Nail
- String
- Water
- Fire Source

**Instructions:**
1. Lay the unopened soda can on its side and poke an angled hole through one side, near the top, with a nail.
2. Turn the can over so that the fluid inside the can will drain out.
3. On the side opposite to the first hole, poke another hole in the can at the opposite angle of the first hole.
4. Let all of the fluid inside of the can drain out and blow into the holes if needed to get all of the liquid out.
5. Submerge the can under water and let it ¼ of it fill up with water
6. Turn the ring of the can normally used to open the can around and bend it up making sure to not open the can.
7. Tie one end of the string to the ring of the can
8. Tie the other end of the string to a an object that allows it to hang freely.
9. Place the fire source underneath the hanging can. As the water inside the can heats up, steam is released through the holes and the apparatus begins to spin like a steam engine based on the principles of Thermodynamics

# R.E. ENGINEERING

## EXPERIMENT 6: Separate Water into Hydrogen and Oxygen via Electrolysis

**WARNING:** This experiment can be dangerous. Proceed with caution! Minors should get the permission and supervision of a parent or guardian prior to performing this experiment. The Author is not liable for accidents that occur while performing these experiments.

**Materials Needed:**
- 2 Clear Plastic Bottles
- 1 Gallon bucket
- 2 Balloons
- 2 metal Hangers
- Tablespoon of Sea Salt or Baking Soda
- 1 9 Volt Battery
- Scissors or Knife
- Connection Leads

**Instructions:**
1. Fill the Bucket with water
2. Cut the bottoms off of the Plastic Bottles
3. Mix Sea Salt or Baking Soda in the water
4. Bend the metal hangers into a "U" shape
5. Place the plastic bottles onto one end of the "U" shaped hangers
6. Place the Balloons over the nozzle of the bottles
7. Connect the ends of the voltage source to the other end of the "U" shaped hangers.
8. Let the device sit until the balloons begin to fill up.
9. Remove the balloon that was over the bottle connected to the positive end of the voltage source; this balloon will be the bigger of the two and will contain Hydrogen gas
10. To test if Hydrogen filled the balloon, release the gas within the balloon over a fire and a small "popping" sound (explosion) should occur

## EXPERIMENT 7: Create "Naphtha" Biodiesel Fuel

**WARNING:** This experiment can be dangerous. Proceed with caution! Minors should get the permission and supervision of a parent or guardian prior to performing this experiment. The Author is not liable for accidents that occur while performing these experiments.

Biodiesel refers to a vegetable oil- or animal fat-based flammable liquid (Naphtha) diesel fuel consisting of long-chain alkyl Acetic Ether (Esters).

### Materials Needed:

- 1 liter Vegetable Oil
- 0.12 oz of Lye (Sodium Hydroxide)
- 6.8 oz of methanol
- Plastic bottle
- Blender (optional)

### Instructions:

1. Pour the Methanol and the Lye into a container and mix them together for 5 minutes until the Lye has completely dissolved. This mixture makes Sodium Methoxide.
2. Pour the vegetable oil into the container with the Methanol and Lye mixture (Sodium Methoxide) and mix the ingredients for 30 minutes.
3. Let the mixture sit undisturbed for 4 hours.
4. After 4 hours, the mixture will have separated into two layers. The top layer if Biodiesel and the bottom layer is Glycerin.
5. Carefully pour the Biodiesel from the top layer into a separate container being sure not to include any Glycerin in the container with the Biodiesel.
6. The Biodiesel you have just made can be used in any Diesel fuel engine.

# R.E. ENGINEERING

## EXPERIMENT 8: Create a Black Light Candle

**WARNING:** This experiment can be dangerous. Proceed with caution! Minors should get the permission and supervision of a parent or guardian prior to performing this experiment. The Author is not liable for accidents that occur while performing these experiments.

**Materials Needed:**

- 1 Cup of "Melt & Pour" Soy candle wax
- Glass bowl
- 20 drops of Black lamp oil
- Stirring stick
- 1 oz of Blue Pigment for Candles
- 10 inch Wick
- 8 oz Candle jar

**Instructions:**

1. Put the "Melt & Pour" Soy candle wax in the glass bowl, and place it in the microwave on high long enough for the wax to completely melt.
2. Use the stirring stick to mix the black lamp oil and the blue candle pigment into the glass bowl of melted wax.
3. Put the 10 inch wick into the 8 oz Candle Jar so that the bottom of the wick touches the bottom of the Jar and let the remaining length of the wick hang over the side of the Jar.
4. Pour the melted candle wax from the glass bowl into the 8 oz Candle jar and let the wax sit in the jar for 5 hours to harden.
5. Cut the wick so that only 0.5 inches of the wick sticks out from the hardened candle wax. Light the Candle and observed the "Black Light" glow of the Black Flame Candle.
6. Additional colored flames can be created by pouring Methanol (sold in hardware stores as HEET antifreeze) over a variety of different elements: Methanol by itself creates a Blue Flame; Potassium Chloride (sold as Salt substitutes in Grocery stores) creates a Violet Flame; Sodium Chloride (sold as Salt in Grocery stores) creates a Yellow Flame; Boric Acid (sold as Roach killer in hardware stores) creates a Green Flame; Calcium Chloride (found in Tofu and Fire extinguishers) creates an Orange Flame; and Strontium Chloride (sold as Sensodyne toothpaste in Grocery stores) creates a Red Flame

## 9 E.T.H.E.R.

### EXPERIMENT 9: Ankh Wireless Energy Transfer Resonant Transformer

**Materials Needed:**
- 40 feet of 22 gauge magnet wire
- 1 LED (Light Emitting Diode)
- Function Generator / Signal Generator capable of 150 kHz 6 V output
- 2 0.02 µF Capacitors
- Electric Tape

**Instructions:**
1. Hold a flame briefly to the magnet wire to burn off the enamel coating
2. Wrap the magnet wire 40 times in coil around a circle of diameter 5 inches and use the electric tape to secure the coil once complete. This will be the primary coil
3. Connect a 0.02 µF Capacitor in parallel to the two ends of the primary coil. (Note: the 0.02 µF Capacitor is only used if you are using a frequency of 150 kHz. If your Function Generator/Signal Generator has a different output frequency, the value of the capacitor should be calculated from the equation $f = \dfrac{1}{2\pi\sqrt{LC}}$. For frequencies below 10 kHz (like the 120 V 60 Hz in most wall outlets), this experiment is not practical
4. Connect the Primary Coil and Capacitor combination to the signal generator, and set the signal generator for 6 V 150 kHz output.
5. Wrap the magnet wire 40 times in coil around a circle of diameter 3.5 inches and use the electric tape to secure the coil once complete. This will be the secondary coil
6. Connect a 0.02 µF Capacitor in parallel to the two ends of the secondary coil.
7. Connect the Secondary Coil and Capacitor combination to the LED.
8. Bring the Secondary coil and capacitor connected to the LED within close proximity to the Primary coil and the LED will light up due to the Resonant Energy transfer.

## 9.0. AFRICAN CREATION ENERGY ETHER

The word "**Physics**" is a Greek word meaning "**Nature**" and has been used to refer to the scientific study of Space, Matter, and Time. In the field of Science called Physics, "**Energy**" is defined as simply "The Ability to do **work**" or the amount of **Work** that can be done by a **Force**. In the field of Science called Physics, a "**Force**" is defined as anything that causes the **Change of Position** and hence suggests **Movement**. Only the field of Science called Physics (Nature) in the area of **Thermodynamics** (which studies the movement and transformation of Energy) can there be found any Scientific Laws (that which is not assumed but rather known and accepted as an undisputable fact). One of the Laws of Physics (Nature) is called "**Conservation of Energy**". The Conservation of Energy Law of Nature (Physics) states that **Energy cannot be created** in the sense that it comes from Nowhere and Nothing into existence, and **Energy cannot be destroyed** in the sense that it ever ceases to exist, but rather **Energy can Transform or change from one form to another**. Energy has many forms including Potential Energy, Kinetic Energy, Atomic Energy, Heat Energy, Electromagnetic Energy, Chemical Energy, Gravitational Energy, Sound Energy, and all phases of Matter. Everything in existence is made up of some form of energy. In Nature (Physics), **Power** is scientifically defined as the rate or amount of time it takes for **work** to be done or **energy** to be converted. In Nature (Physics), **Radiation** is scientifically defined as any process by which the energy emitted by one object travels through a medium and is eventually absorbed by another object. In the field of Science called Physics (Nature), a **Black Body** is scientifically defined as an ideal object that absorbs all electromagnetic radiation (Light Energy) that it receives.

## 9 E.T.H.E.R.

**Creation** refers to the act or process of causing something new or novel (nova) to exist or come into being. As it follows from the discussion about Energy, since everything in Existence is composed of Energy, and Energy can only change from one form to another, then Creation and Creating does not mean bringing about something new into existence from Nothing and Nowhere, but rather Transforming or changing the form of one thing, or combining the forms of several things, into the desired new thing. Creation of anything in any form is a gradual process of **Growth** and **Change** over time in which a metaphorical or literal seed, nut, kernel, or node grows and transforms into another form, figure or structure. A new creation is always the initiation or beginning of one thing and the termination or ending of another thing. Change, Growth and Decay are mathematical concepts which indicate the increase or decrease of a quantity, magnitude, or multitude over time respectively. Creativity is the level or degree of creative mental ability, and a creation always exists as a thought, idea, or concept in the mind of its creator before manifesting in the physical world. Creation occurs through Creativity, and one of the most important Creativity techniques is "Problem Solving". Problem Solving occurs when it is desired to go from one state to another state (change of position), and therefore from the scientific definitions given, "Problem Solving" and **Creativity** are both a mental **Force** which can be measured as **Energy**.

From the scientific definitions given, it follows that **African Creation Energy** can scientifically be defined as the Work, Effort, Endeavors, and Activities of Black or African people that cause a movement or change. African Creation Energy is The Energy, Power, and Force that created African people and that

African people in turn use to Create. Since African people are the Original people on the planet Earth, it follows from thermodynamics that the Creation Energy of African people is the closest creation Energy of all the people on the Planet to the **Original Creative Energies** that created the Planets, stars, and the Universe. **African Creation Energy** is **Black Power** in the scientific sense of the word "Power", and this book **Radiates** African Creation Energy to be absorbed by the **Black Body**.

Amongst the **Yoruba** people in **Nigeria**, **African Creation Energy** is called **"ASHE"** which means "the power to make things happen" and the various forms of African Creation Energy are personified in the Yoruba "Orisha" (Ori-Ashe) deities. Amongst the **Akan** people in Ghana in the Twi language, **African Creation Energy** is called **"TUMI"** – the web of energy and power that exists throughout space and all of creation which was woven and designed by the Akan deity of wisdom Ananse Kokuroko. In the Congo, **African Creation Energy** is called **"DIKENGA"** which refers to the energy of the universe and the force of all existence and creation, and the thermodynamic process of the transformation of energy is depicted in the Congo cosmogram called the "Yowa". Amongst the **Mande** people of West Africa, the Creative Force called **Nyama** or **Amma** is used by the Blacksmiths as a means to forge technology for the well-being of the entire village. In Ancient Africa, Egypt, Nubia, KMT, Kush, Tamare, *etc. et al*, **African Creation Energy** was called **Sekhem** which was energy that individuals used to control the elements of nature to create anything desired. In books written by Afroo Oonoo, the African Original Creation Powers of The Universes are called **NoopooH**, and in books written by Amunnub-Reakh-Ptah, the original creative force of African people is called **Nuwaupu**.

## 9 E.T.H.E.R.

The conduit of "African Creation Energy" who has written and authored this book, and other books, goes by the title of **Osiadan Borebore Oboadee** from the Twi language spoken in Ghana West Africa. The Twi word **"Osiadan"** comes from the root words "Si" meaning "Build" and "adan" meaning "Building" with "O-" being a way to denote a "Master". Hence "Osiadan" literally describes a **"Master Builder of Buildings"**. Also note the phonetic similarities between the Twi words "Si" and "Adan" and the Ancient Egyptian words **"Sia"** (wisdom) and **"Aton"** (high noon sun). The Twi word **"Borebore"** comes from the root words "Bo" meaning "Create" and "Re" meaning "to do repetitiously", thus "Borebore" is used to describe a **"Continuous Creation"** or **"Architect"**. The word "BoreBore" or "Bore" in Twi is also related to the Hebrew word **"Bara"** meaning **"to begin"** found in the first verse of the first chapter of the Judeo-Christian Bible, and is also related to the Yoruba word "bere" meaning "to begin". Also note the phonetic similarities between the Twi words "Bo" and "Re" and the Ancient Egyptian words "Ba" (soul) and "Re" (sun). The Twi word **"Oboadee"** comes from the root words "Bo" meaning "Create" and "Abode" meaning "Creation" with "O-" being a way to denote a "Master", hence "Oboadee" literally describes a **"Master Creator of Creations"**. Oboadee is also pronounced O-Poatee in different African dialects, and is said to derive from the pronunciation of the name of the Ancient African Creation deity PTAH. Osiadan, Borebore, and Oboadee are three principles of **African Creation Energy**. Osiadan Borebore Oboadee is African by blood and lineage; a descendant of the **Balanta-Bassa** and **Djola-Ajamatu** tribes in present day **Guinea-Bissau (Ghana-Bassa)** West Africa. Both the Balanta and Djola tribes migrated to West Africa in Ancient times from the area which is present day **Egypt**, **Sudan**, and **Ethiopia**. Osiadan Borebore Oboadee is a descendant of the Ancient Napatan, Meroe, Kushite Pyramid Builders, and a Scientist,

Engineer, Mathematician, Problem Solver, Analyst, Synthesizer, Artist, Craftsman, and Technologist by education, profession, and nature. Osiadan Borebore Oboadee has obtained Bachelors and Masters Degrees in the areas of Electrical Engineering, Physics, and Mathematics between the years of 2003 and 2006. Born into the African Diaspora, Osiadan Borebore Oboadee made his first trip to the African continent in the year 2008. Between the years of 2009 and 2010, Osiadan Borebore Oboadee set out to develop, engineer, invent, formulate, build, construct, and create several Technologies (Applications of Knowledge) for the well being of African people worldwide and attempted to radiate the energy that motivated and inspired the development of those technologies in a three part introductory educational series which collectively was entitled "The African Liberation Science, Math, and Technology Project" **(The African Liberation S.M.A.T. Project)**. The three books that are part of African creation Energy's "African Liberation S.M.A.T. project" are:

1. **SCIENCE:** (Knowledge/Information)
   The SCIENCE of Sciences, and The SCIENCE in Sciences
2. **MATHEMATICS:** (Understanding/Comprehension)

   $9^{9^9}$ Supreme Mathematic African Ma'at Magic
3. **TECHNOLOGY:** (Wisdom/Application)
   P.T.A.H. Technology: Engineering Applications of African Science

Osiadan Borebore Oboadee's primary purpose for writing the books of the "African Liberation S.M.A.T. Project" was to motivate the Creative Energies, Minds, and Bodies of African people to go from an inert state of Theory and Speculation to an Active creative state of Development, Creation, and Productivity for the survival and well-being African people everywhere. It is the goal of African Creation Energy's "African Liberation S.M.A.T. project" to free the minds, energies, and bodies of African people from mental captivity and physical

## 9 E.T.H.E.R.

reliance and dependence on inventions and technologies that were not developed or created by, of, and for African people. In 2011, at the age of 30, after writing the books of the "African Liberation S.M.A.T. Project", Osiadan Borebore Oboadee found it necessary to provide evidence of the African Creation Energy Philosophy in Action and Application by building structures and thus embarked upon the project of building a Pyramid and authoring a text entitled **"ARCH I TET: How to Build A Pyramid"** as part of his **30 year "Djed Festival"** of renewal.

Following the multitude and plethora of information presented by the many Scholars (who have affectionately been labeled **"MASTER TEACHERS"**) who have came to improve the conditions of African people, it is the goal of **Osiadan BoreBore Oboadee** and **African Creation Energy** to be the catalyst in the synthesis, unification, and practical application of the information presented by the great Master Teachers. Thus, it is the aspiration of Osiadan Borebore Oboadee and "African Creation Energy" to be and breed **"Master Technicians"** who TeAch through Action and Application. In order to achieve this transformation, Osiadan Borebore Oboadee has strategically cultivated African Creation Energy in accordance to an Ancient African Alchemical formula. The chosen term "African Creation Energy" has hidden "Etheric" meaning. The word **"Africa"** comes from the Afro-asiatic word **"Afar"** meaning **"dust"** which represents the **"Earth"**. The etymology of the word **"Creation"** comes from the word **"Crescent"** which represents the **"Moon"**, and the word **"Energy"** represents the **"Sun"** which is a primary source of energy to our planet. Therefore, one of the hidden meanings in the coined term **"African Creation Energy"** represents the **"Earth, Moon, and Sun"** cosmic forces also known **"Ptah, Aah, and Re"** or **"Re Ah Ptah"** or **"Space, Matter, and Time"** in Ancient African Cosmology.

www.AfricanCreationEnergy.com

Moreover, the abbreviation of **"African Creation Energy"** is A.C.E. which spells the word "Ace" which has etymological meanings of "a unit, whole, one, first, one who excels, and Primary" and indeed African Creation Energy represents the excellent Primary and **Original Creative Forces** in Nature.

The letters used to abbreviate "African Creation Energy", A.C.E. not only spell the English word "Ace" meaning "First, Primary, or Original", but also represent the 3 fundamental geometric shapes of the **Triangle** represented by the letter **"A"**, the **Circle** represented by the letter "C", and the **Square** represented by the letter **"E"** which when combined form the Ancient **Alchemical symbol** of the **"Squared Circle"**. Utilizing African Creation Energy, Osiadan Borebore Oboadee has used the "Squared Circle" as a metaphor and symbol to represent the unification of dualities necessary to bring about the birth and creation of a new paradigm. Also, all of the books authored by Osiadan Borebore Oboadee and African Creation Energy have been released on specific strategic dates. Each of the four previously released books represents a different element necessary for transformation in Ancient African Alchemy, with this fifth book being released in this pivotal and transitional year during the **Summer solstice of the year 2012** representing the fifth and quintessential Alchemical element of **Ether**.

**A.C.E.**
**African Creation Energy**
**Alchemical "Squared Circle"**

# 9 E.T.H.E.R.

## African Creation Energy Books

| # | BOOK | TITLE | ELEMENT |
|---|------|-------|---------|
| 1 | | The SCIENCE of Sciences and The SCIENCE in Sciences<br><br>Release Date: 10-10-10 | Air |
| 2 | | $9^{9^9}$ Supreme Mathematic, African Ma'at Magic<br><br>Release Date: 09-09-09 | Water |
| 3 | | P.T.A.H. Technology: Engineering Applications of African Sciences<br><br>Release Date: 05-04-10 | Fire |
| 4 | | ARCH I TECT: How to Build A Pyramid<br><br>Release Date: 11-11-11 | Earth |
| 5 | | 9 E.T.H.E.R. R.E. Engineering<br><br>Release Date: 06-26-12 | Ether |

www.AfricanCreationEnergy.com

# R.E. ENGINEERING

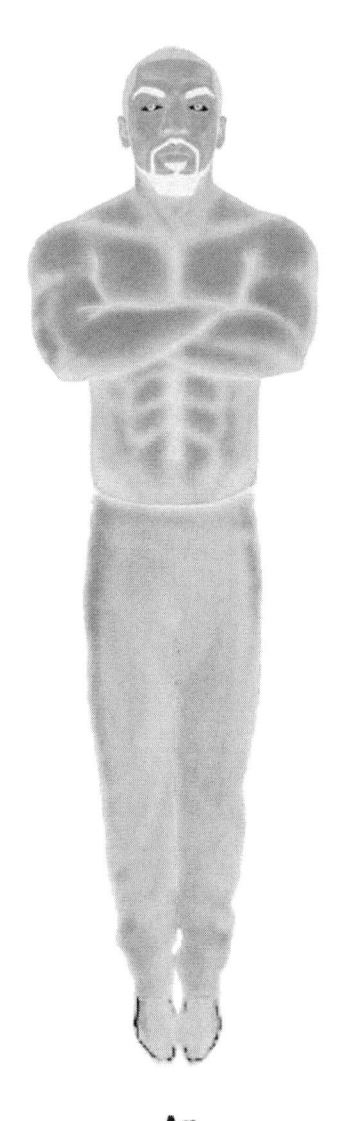

An
Æther Bender
"Prophessor A.C.E."
Osiadan Borebore Oboadee
www.AfricanCreationEnergy.com

## 9 E.T.H.E.R.

# **RE**FERENCES AND **RE**SOURCES

1. "Introduction to the Nature of Nature Book 1" by Afroo Oonoo

2. "Moonset and Sunrise in the Nature of Nature" by Afroo Oonoo

3. "P.T.A.H. Technology: Engineering Applications of African Sciences" By African Creation Energy

4. "Stolen Legacy" by George G.M. James

5. "The Holy Tablets, Chapter One: The Creation, Tablet One: Epic of Creation and Before" by Malachi York

6. "The Man from Planet Rizq" by Dr. Malachi Z. York

7. "The Nine Ball Count I & II Liberation Information" by Wu-Nupu, Asu Nupu, and Naba Nupu

8. "The Sacred Records of Atum-Re - The Black Book part 2" by Amunnubi Ruakhptah

9. "What is Spirit and Soul" by Dr. Malachi Z. York

Printed in Great Britain
by Amazon